PENGUIN BOOKS

Headlines and Hedgerows

John Craven was born in Leeds and started his journalistic career as a junior reporter on the *Harrogate Advertiser.*

In 1972 he launched the world's first television news bulletin for children, *John Craven's Newsround*. In 1989, after 3,000 episodes, John left to present the steadfast television institution that is *Countryfile* and has brought the glorious British countryside to our television screens ever since.

Headlines and Hedgerows

JOHN CRAVEN

PENGUIN BOOKS

PENGUIN BOOKS

UK | USA | Canada | Ireland | Australia
India | New Zealand | South Africa

Penguin Books is part of the Penguin Random House group of companies whose addresses can be found at global.penguinrandomhouse.com.

First published by Michael Joseph 2019
Published in Penguin Books 2020
001

Set in 12.01/14.33 pt Garamond MT Std
Typeset by Jouve (UK), Milton Keynes
Printed and bound in Great Britain by Clays Ltd, Elcograf S.p.A.

A CIP catalogue record for this book is available from the British Library

ISBN: 978–1–405–93269–1

www.greenpenguin.co.uk

To my family

Contents

Prologue

The microphone was on and I was halfway through reading the news when the door opened and in came my Aunty Ethel with some shopping. Then a member of the audience opened the other door and the two of them began talking about what was in the bags while I struggled to continue the bulletin.

This unlikely scene was happening not, I'm glad to say, in a studio, but in a kitchen in suburban Leeds. The 'newsreader' was me in my early teens and the 'audience' was my mum, who had been listening patiently to my early broadcasting efforts. The microphone was a birthday present I had wired from the kitchen table to the radio in the sitting room and my 'news script' was the front page of the evening newspaper.

Pretty soon Mum and Aunty Ethel realized I was far from pleased they had broken into the 'bulletin' so they made a pot of tea and sat down to listen to the rest of my news. Fortunately the interruption did not put me off a career in broadcasting (though it did take me some years to realize that was what I really wanted to do). I have talked into microphones on BBC One almost every week for more than fifty years, and occasionally on BBC Two and ITV as well.

During that time I have been involved with three groundbreaking programmes – *Newsround*, *Swap Shop* and *Countryfile* – of which I am enormously proud, worked with some great teams who have been friends as well as colleagues, travelled to more than ninety countries and met a vast number of people from prime ministers to pop stars, from inquisitive children to a future saint. I count myself tremendously fortunate to have survived so long in an industry which can be notoriously fickle but which I love unconditionally.

We have all heard that well-known piece of advice first coined by W.C. Fields: 'Never work with animals or children'. Well, I have done both throughout my career (in fact, I couldn't have succeeded without them!) so in my case at least that old adage is totally wrong. I suppose one reason for my longevity is that I have never been very ambitious. I have not sought the headlines, never seriously courted celebrity nor been tempted to take chances on high-profile but potentially risky and short-lived programmes – apart from one, and that was *Newsround*, which was a six-week experiment in 1972. Thankfully it is still going strong so, as it turned out, it was not much of a gamble and a recent poll in *Radio Times* placed *Newsround* at number three in a list of the top twenty children's programmes of all time.

And *Countryfile* is often in the top twenty of most-watched shows. During my thirty years there I have seen rural issues ranging from social isolation and deprivation to the way our food is produced climb higher and higher up the national agenda, making the programme a vital conduit of information to millions of concerned viewers. The fact that our audience is split pretty evenly between country dwellers and townies proves to me that, united as a nation in this at least, we want to preserve, protect and enjoy our glorious countryside. Every week, we highlight the problems it faces (in my lifetime half our hedgerows have gone and so have two thirds of our small family farms) while at the same time showcasing the life-enhancing experiences it can offer. This dichotomy is what makes working on *Countryfile* so rewarding.

I have always been content to be a jobbing broadcaster, taking each day as it comes and hoping it will bring good stories and interesting people – which is probably why I have never been out of work in half a century. Journalism has taken me from the cobbled streets of my childhood to the country roads that crisscross our rural heartland, and most of that time it has been by way of the world's greatest broadcasting organization, the BBC.

On my first attempt to film a news story for television the cameraman told me: 'A movie camera either likes you or not and John, you're lucky.' So, being a firm believer in making your own luck, I stuck with broadcasting. But my first love was the printed word and I thought it was time to put down some of my lifelong experiences in this book, while I can still remember!

My watchword has always been: keep it short and keep it simple without being simplistic. Hopefully the latter still applies but the former has gone out of the window. I hope you enjoy wandering with me down a memory lane that I am so glad, so grateful to have trod.

Oxfordshire, 2019

1. Dad

A beaming five-year-old rides on the shoulders of a stranger, a hero returning from fighting tigers in distant lands. For weeks the boy has longed for this moment because the man is the father he could not remember, who left home for life in the jungles when the boy was just a few months old.

When the man had stepped from the train onto the steam-shrouded platform at Leeds City station, the boy's mother, aunts and grandmothers had rushed to the stranger and engulfed him with love and tears. Then the man had picked up the boy, embraced him long and hard and placed him on his shoulders.

But how can this little boy understand the reality of the situation; that his hero, emaciated and worn out by illness and ill treatment, barely has the strength to lift him the short distance to the taxi that will carry them home. The jungles were real enough but the tigers had been living only in the boy's imagination. What his father has fought, and miraculously survived, is the monstrous inhumanity of an enemy who has held him captive for three and a half years.

That stranger was my dad, Private Willie (it should have been William but his parents panicked at the font, so he was known forever after as Bill) Craven, coming back to his native Yorkshire with no thanks at all to the Imperial Japanese Army. He had heard whispers his ordeal was over on 16 August 1945 – my fifth birthday.

In the taxi our little family – my mother Marie (pronounced Marry), Dad and I – were together again, despite all the odds, for the first time in four years. For most of that time my mother had lived in hope, with me as a reminder of their love, because she had no idea whether he was alive or dead. He had disappeared into a hell-hole thousands of miles from home and her many letters to him went unanswered. Now, that hope had been fulfilled.

Dad had a conventional working-class upbringing in Leeds until the Second World War broke out. Two years earlier he had married his childhood sweetheart and been promoted to manager of the Ideal Dividend grocery store on Kirkstall Road. I was born on the busiest day of the Battle of Britain and two months later Dad enlisted with the Royal Army Ordnance Corps as a motor mechanic.

A few months after that he was posted to Malaya, never before having left the confines of Yorkshire. What an experience it must have been, what a contrast to the life he had known – being merged into the military system, then the long sea voyage with the risk from enemy submarines, followed by the tropical heat and tension of Singapore in its final days as a British stronghold in Asia.

At least he had the chance to get acclimatized to the conditions before the Japanese invaded. Troops who arrived to strengthen the 'fortress' shortly before the end dropped like flies. It was from there he sent me a telegram on my first birthday, the last time I was to hear from him until the war was over. The message said simply 'Loving Birthday Greetings. I wish we were together on this special occasion. Fondest love and kisses. Daddy Craven.'

When Singapore fell six months later, Winston Churchill

described it as 'the biggest capitulation in British military history'. My dad was one of 80,000 British, Australian and Indian troops taken prisoner when the victorious Japanese completed their occupation of this outpost of the Empire at the tip of what was then Malaya.

The island's big guns had been pointing their armour-plated shells out to sea, expecting a naval assault, but instead the enemy attacked from the land. Although many of the guns did manage to turn towards them, their shells were intended to destroy ships, not moving soldiers, and the Japanese swarmed in.

The only thing my dad ever told me about the battle for Singapore was that, while most of his unit tried to escape to Java, he and a few others were ordered to stay behind. They were told to disable vehicles and destroy whatever ordnance they could to prevent the Japanese from making use of them. With bombs exploding all around he drove lorries into the harbour, then they had a few beers and waited for the inevitable. What followed was three and a half years of hell, the first eight months of it spent in Singapore's Changi Jail, the last twenty-one months or so in prisoner-of-war camps in Thailand, and the thirteen months in between slaving on the infamous Death Railway.

The line ran for nearly 260 miles through inhospitable jungle terrain to carry supplies from Thailand to Japanese troops in Burma (now Myanmar), and every inch of it was built by forced labour. As many as 100,000 local South East Asian workers, known as coolies, died laying rails and sleepers and building several hundred bridges – as did nearly 13,000 Allied prisoners of war. With no medication, sparse food, little rest and tropical diseases scything through the weakened workforce, the railway truly lived up to its awful nickname.

I know from the War Office questionnaire my dad had to complete after he was freed that on 3 November 1942 he was among the first to travel north from Singapore by train, crammed into locked metal goods wagons for three days and nights with no rest stops, no latrines and little food. The heat and smell was appalling and those who died were thrown out of the wagons by the guards. When they arrived at what they had been led to believe would be a Red Cross camp they discovered it was in fact Won Rum, the first of three work camps that Dad would be posted to along the Death Railway.

Like many of his compatriots who eventually came home, he refused to talk in any detail about those times, no matter how much I tried to persuade him both as a son and as a journalist seeking the truth.

There was a conspiracy of silence, a compact of secrecy, among many of the survivors, and when it was broken in books such as *The Naked Island* by Russell Braddon, and most dramatically by the film *The Bridge on the River Kwai*, my dad felt outraged and betrayed. When he received an invitation to the film's premiere, he turned it down.

The reason for this silence, I have always suspected, is that he and thousands of others suffered beyond our comprehension at the hands of their brutal captors, who held the belief it was better to die than be taken prisoner. By refusing to accept such a creed, and in the struggle to save themselves and their fellow captives, in the eyes of the Japanese they had compromised their standards as men.

Over the years I have seen many documents, official and

otherwise, covering this black chapter in the history of war. Here is an excerpt from one written by Lieutenant Colonel G. E. Swinton, who was the senior British officer at one of the camps my father had the misfortune to be in:

> Regardless of the cost of human life and suffering the construction of the railway had to proceed at top speed under all conditions of weather. The result was the most ghastly and inhuman treatment of prisoners.*
>
> The engineers insisted on the strength of the working parties being maintained at a level far beyond the availability of fit men with the result that large numbers of men suffering from chronic diarrhoea, malaria, beriberi and ulcers were forced out to work under the most appalling weather conditions and often for twelve hours daily.
>
> If the numbers of men whom the British MOs considered by the wildest stretch of the imagination fit for work were not sufficient – and they hardly ever were – the camp authorities paraded all the sick men and ordered a further number out. Further, even worse than this, the (Japanese) engineer officer frequently came into camp and, after holding his own sick parade, turned further men out.
>
> Unfit men were divided into sick; light sick; light, light sick; but on many occasions their work was the same. As to their treatment while at work it was in some cases brutal to a degree. One sergeant in particular took a delight in hitting men on the head with an iron rod. Men were often struck

* The guards used to yell 'Speedo! Speedo!' when orders went out to up the pace of construction. It was a word which, years later, my dad would call out to me with some irony if I was dallying!

with the fist or with bamboos, while the engineer officer actually seemed to encourage his men to treat prisoners brutally.

After reading such accounts, I wonder how on earth my dad – always slightly built and smallish – managed to survive. It must, to be honest, have been down to good luck and a big helping of Yorkshire grit, but survive he did and, like a creaking gate, lived until he was eighty. But that pact of secrecy stayed with him to the end. He told me only a handful of stories from his long internment. Like the night he collapsed from exhaustion by the side of the railway line after laying sleepers all day near the Kwai bridge and hit his head on what he thought was a rock. Daylight revealed that his 'pillow' had been an unexploded bomb!

After the Death Railway was completed he spent nearly two years in Tamakan, a so-called 'hospital camp', although it had no medical supplies or facilities. Not the best place to develop appendicitis, which Dad did, and it threatened to turn into peritonitis and hasten his end. Fortunately, one of his fellow prisoners was a Harley Street surgeon and, despite the lack of any of the equipment, he decided to operate.

Amongst his crude instruments were knives made from sharpened bamboo and for an anaesthetic/antiseptic the surgeon used saki, a potent rice wine secretly brewed by the POWs. Every day they donated a few grains of rice from their meagre food ration to a 'saki fund' for such emergencies. Dad was so weak that after a few swigs of saki he was out cold and the surgeon got to work, so successfully that, despite the primitive 'operating theatre', his patient made a complete recovery.

Years later, when he was safely home, a letter arrived for Dad from the surgeon with a bill for £50 and a lovely letter saying that would have been the charge in Civvy Street. Thankfully, though, not in Tamakan.

POWs in the camp had to constantly repair the bridge over the River Kwai after it had been targeted by Allied bombing. Dad said they were not allowed to move from their positions on the bridge until the Japanese gave the order, putting them at real risk of being killed by the RAF. Weeks before Japan's surrender, he was forced to join a work party sent north to repair the new railway line following bombing by the Allies and encroachment by the jungle. From there he was moved on to Longshi in Burma and ordered to hide Japanese supplies in caves ahead of the Allied advance.

It was hard labour for weak and hungry men but thankfully the ordeal did not last long. It was there, on the very day that I was five, that he heard unofficially from local workers that the war was over. But could it be just a rumour? The POWs knew for certain ten days later when the RAF dropped leaflets, not bombs. On 5 September he was allocated four sheets of paper and a pencil and wrote his first letter home:

Dear Marie. At last I am able to write a letter and I hardly know where to start. It has been a terrible time but I have been lucky and got through with the help of God and your love and prayers.

I got your first letters after being taken POW at the back end of May 1943 (four yours, two from Mother) at a camp which was one of the worst on the railway, which we knew as 211. I was suffering from dysentery at the time.

The latest letter I had from you was written on July 10 1944 (with three photos of John). I can hardly realize I have a boy as old as he is.

When I left home he was only a baby though I am glad that he came along and I have great plans for him when I come home. I missed his baby years but I will make up for that in the future.

We arrived at this camp yesterday from further north; the Japs brought us down and handed us over to a British officer in charge of the area. We don't know what the plans for the near future are but we are hoping that it won't be long before we are on our way home. It is what we all want.

We little thought when I left home that it would be all this time but I love you more than ever and am longing for the day when I can take you in my arms again. These lonely years have been a great strain, more for you I think than for me, but I am certain that our love has stood it. The pencil I am using four of us have managed to borrow and are waiting, one after the other, to use it.

What I want is a really good English dinner, Yorkshire pudding, roast beef and all the trimmings. It took me a long time to get used to rice and I will not be sorry to leave it. I haven't had any bread made with wheat flour for over three years.

Well, darling, space is getting short so cheerio and God bless you. All my love. Yours forever. Bill.
To John. A little message to say be good to your mammie always and look after her until I come home. Love from your Dad.

Back in Leeds, that letter brought an end to years of suspense and heartache for my mother – the agony of not knowing whether Dad was alive and, if so, what state he was in. She had been lovingly supported by my grandmothers – both my grandfathers were dead – and half a dozen uncles and aunties, so I must have been spoilt rotten. Certainly, I was the first kid in the neighbourhood to get a tricycle – a rarity because many had been melted down for the war effort.

Every bedtime Mum used to sing to me a song that had been one of their favourites – 'You Are My Sunshine' – and we would imagine Dad singing it along with us somewhere in the jungle.

> The other night, dear, as I lay dreaming
> I dreamt that you were by my side
> When I awoke, dear, I was mistaken
> So I hung my head and I cried
> You are my sunshine, my only sunshine
> You make me happy when skies are grey
> You never know, dear, how much I love you
> Please don't take my sunshine away.

And when the war ended in Europe, and the country went mad with joy on VE-Day, one four-year-old boy at a celebration bonfire party in Grimthorpe Street, Headingley, was very confused. Why all this happiness when my dad was still far away in a jungle? Why hadn't he come home? Unknown to me, the fighting was still going on in the Far East and with all the celebrations in Europe it was becoming known as the Forgotten War.

Not until the Americans dropped their atomic bombs on Hiroshima and Nagasaki was it to end and, horrific though the consequences were, that ultimate act of war freed my father and many thousands of his fellow prisoners and allowed them to return home.

Years later, Dad wrote this account of his homecoming:

> It was a weekend I shall never forget. I had arrived back in Liverpool, which I had left four and a half years before

when I was called up for the Armed Forces. My son John was only six months old when I got my five days embarkation leave in March 1941.

There was a dock strike on at the time but the dockers came in to get us safely tied up at the quay. Just before lunch we were allowed to disembark and we were taken to an Army camp a short distance away.

It was about eleven o'clock at night before we were seen by various doctors, they were mostly interested in getting the POWs away as soon as possible.*

I weighed five and a half stones at the time and one of the doctors gave me a chit so I could get six eggs (one week only) to build me up! The following morning we were taken to the station where I met a few men from my unit that I had not seen since the fall of Singapore. My brother had come to Liverpool to meet me so we got the train to Leeds.

We arrived at the City station about 12.30 p.m. and there at the end of the platform was Marie and John and most of my relations. John came to me as if I had been away overnight. It was as I hoped it would be because I had heard many dads had been rejected by their children when they had been away for so long.

There was much hugging, laughing and tears, etc. but John was not going to let me go again. Eventually Marie, John and I got a cab home; well I say cab but it was a small Austin 8. The driver was a very kind man; he said he was so pleased to take us home he would not take the fare.

* This in spite of the fact that Dad had suffered from malaria, beriberi, dysentery, diarrhoea and ulcers.

When we got to the house there were flags flying in the street and a big flag over the door with 'Welcome Home' written on it. Marie gave me the key to open the door. When we were inside John wanted to see what I had brought with me in my bag. I had packed as many things as I could afford on the amount of money that I had been allowed; mostly sweets, chocolates, oranges, apples, grapes and a bunch of bananas.

John had not seen a banana before so we broke one off and he tried to eat it with the skin on. We all laughed again and showed him how to take the skin off and eat it; he really enjoyed the taste. The next morning Marie was up early to prepare the breakfast so John came into our bed as a special treat. Many friends, relations and neighbours had brought gifts of food out of their small rations so we were well fed for a few days.

In the afternoon some people from Halifax came to see me. I did not know them but their son was in the same unit as me but he had not returned. They wondered if I knew anything about him. We talked for a long time and I said I was so sorry I could not help them as I had not seen him since about two weeks before Singapore fell. As far as I knew he had gone to Java with the rest of the unit. Twenty of us had been left in Singapore to demobilize a vehicle park.

After that they went home and we three were left alone again with our memories. It seemed like heaven after being in hell all that time.

There was an ironic parallel between Dad and his father, William, who served as a sapper in the Royal Engineers during the First World War.

William was something of a mystery man and his name

was rarely mentioned within our wider family, but he saw action at Gallipoli and later at Basra in what was then Mesopotamia but is now Iraq.

Dad was just a toddler when his father joined up. It's not clear what happened to him because his service record was, like so many others from that war, destroyed when warehouses where they were stored in London's docklands were bombed during the Second World War. What we do know is that there were high casualty rates, mainly due to disease as the sappers toiled in heat and dreadful conditions to clear and build waterways for advancing British forces.

My grandfather was one of 377 sick or injured troops being taken from Basra to Bombay in the hospital ship *Varsova*. He was suffering from trench sores and so, like Dad, he was on a slow voyage home by sea to recover. Unlike Dad, he didn't make it and his name is engraved on the war memorial in Basra. On 30 August 1917 he died, so the records say, from the effects of heat and was buried at sea in the Persian Gulf. I didn't discover this information until after Dad's death but I wonder if that connection ever crossed his mind as he faced an unknown fate in the heat and horror of the Death Railway.

2. The Grimthorpe Street Gazette

It took me a while to get used to Dad around the house after being a 'mummy's boy' for so long. But very slowly, very gently, he extended that bond between mother and son to include himself so I never felt there was a usurper in the house. He went back to his old job managing the grocery store but because of what he had endured in the Far East he couldn't stand behind the counter for long periods of time without feeling acute pain. His Japanese captors had taken away not only his health but also his promising career, although he would say he was one of the fortunate ones because they failed to take away his life.

Eventually he took a sitting-down job, constructing delicate components in a factory making water meters. His small team of colleagues became really good friends, and although his ambitions had been lowered his always-delicate health was not unduly tested. My mum was amazing – nursing him when malaria returned, with its fearsome bouts of shivering and shaking, and ensuring that her man would never again be without her loving care.

In fact, Dad suffered from no fewer than twenty-seven bouts of malaria, and so weak was his digestive system as a result of malnutrition in the camps that he had to undergo major abdominal surgery at Liverpool's Hospital for Tropical Diseases. If ever anyone deserved a war pension it was him, but even though my sister, Jean, battled with the Ministry of

Pensions for many years – and she is good at tackling bureaucracy – his case was refused.

Not until the launch in 1975 of a national campaign to draw attention to the disgraceful way that former Far East prisoners of war were being treated by successive governments did anything happen. A distinguished QC agreed to act pro bono on their behalf: Jean sent him Dad's details and he used them, along with the case histories of many other POWs, to secure pensions through the courts.

In 1990 – that's forty-five years after he came home – Dad finally got his pension, back-dated to 1952 but with many deductions which in the first few years were converted from pounds, shillings and pence into decimal currency. It was not a large amount but he felt that at last his suffering for his country had been recognized. But it came too late to provide any benefit. Mum had died a few months before it was awarded and Dad died a few months afterwards.

In November 2000 the British government announced it would pay £10,000 each to surviving prisoners of war of the Japanese 'in recognition of the unique circumstances of their captivity'. Ten years after Dad's death.

For some years after his return from the war our home was in Grimthorpe Street, a largish back-to-back like many others in the terraced rows that sprouted from Ash Road in the residential suburb of Headingley.

Then it was an area for young families and the elderly, but over the years its personality has changed and when I went back recently the cobbled streets had been tarmacked and many of the residents were students.

My mum was the kind of mum that every child should

have; loving and protective and always there when she was needed most. I can't remember her losing her temper but she was no pushover. When she was a teenager she had signed the temperance pledge and didn't touch a drop of alcohol until I was in my thirties, when she decided to tear up the pledge.

So when we went occasionally as a family to the local she would start with a Tia Maria, then try a sherry, then perhaps a half of lager because she was thirsty and so on – switching drinks to sample the different tastes. It never had any effect on her even if the rest us were feeling a little tipsy.

Mum was renowned for her malapropisms. Once when we were driving past a field of cattle she said; 'Look, there's a herd of freesias.' And, when trying on a new cardigan, she asked: 'Is it made of polystyrene?'

She was a bright, practical, intelligent woman who devoted her life to her family and she was taken from us in the cruellest way by motor neurone disease. It began in her late seventies when she was living with Dad in a nursing home in Harrogate. In her final days she was paralysed and could only communicate by blinking her eyes. Her brain was trapped in a functionless body – a terrible fate for anyone, especially my lovely mum. For Dad, she was the love of his life and, after all he had been through, he survived for less than a year without her.

Back in the happier times of my early childhood, Headingley was thought to be one of the better suburbs, a couple of miles from the centre of Leeds with an intriguing mix of styles and people; Victorian and Edwardian houses, smart semis and detached residences from the 1930s, the sprawling Beckett Park, where I learned to play tennis, climb trees and first discovered nature and, of course, the famous Headingley sports stadium, still the home of Yorkshire County Cricket Club and Leeds rugby league team.

Headingley has a history dating back to the Domesday Book and it was a good place to grow up. Anyone who has read Alan Bennett's diaries will know all about it. He was in the senior sixth form of my school when I was a new boy there. We have never met but I remember his father's butcher's shop in Far Headingley, a smaller and rather more select suburb halfway between my home and school. The shop was just round the corner from my grandmother's small cottage and I often went shopping for her there.

With Dad still recovering from his ordeal in the tropics, Britain in 1947 was hit by the coldest winter in three centuries. Roads and railways were blocked by snow for weeks, there was pack ice in the English Channel and power cuts and coal shortages demoralized a nation which had been hoping for better times. All of which my little friends and I were blissfully unaware of as we played for weeks on end in the white blanket that transformed our neighbourhood.

So much snow piled onto the verges on Grimthorpe Street as a result of it being cleared from pavements and the road that it towered above our heads. We dug our way into it, with some adult help, and created our own igloo den that lasted from late January till March. We were the Headingley Eskimos and we loved it.

Eventually the snow melted and the summer turned into one of the hottest on record. In the midst of it, on my seventh birthday on 16 August, my life changed dramatically.

I think I knew that my mother was pregnant but the day started just as I had hoped it would with lots of cards and hugs from aunties and my big present – a lop-eared rabbit in a cage. What I wasn't expecting was for my mother to be then rushed to hospital. A neighbour opposite had the street's only telephone and I waited with her until Dad phoned with the news that I had a sister.

All very exciting, but slowly it dawned on me that this baby was taking all the attention away from me and my birthday and that from now on I would have to share it with her! We'd be twins, seven years apart – and I was not happy about that. My lovely Aunty Ethel gave me a big kiss and said: 'What have you got for your birthday, you lucky boy?' 'A black-and-white rabbit,' I said petulantly.

Looking back, my less-than-brotherly behaviour is unsurprising, because I had been the centre of attention during those long years my dad was missing and in the many months since his return. I was in danger of turning into a rather spoilt brat, if I wasn't one already, so Jean's arrival was a blessing which made me come to terms – albeit reluctantly at first – with the fact that our family wasn't just about me. Slowly I forgave my baby sister for what I'd regarded as her untimely appearance and we became very close. And what possible excuse could we have for forgetting each other's birthday?

My family background is built on hardworking Yorkshire stock and when researchers from BBC One's genealogy show *Who Do You Think You Are?* looked into it they decided it was not interesting enough to fill a programme. But they kindly sent me the information they had dug up and I found it fascinating. I discovered that for generations the men on my father's side had been ironworkers and that my great-grandfather Josh Craven drowned in the Leeds to Liverpool canal in the early hours of an icy, snowy morning on his way to work. He must have lost his footing on the towpath. His widow, Eliza, had, to my surprise, been born in London – I thought all my ancestors were true Tykes.

My paternal grandmother Sarah had been widowed since the First World War and, for reasons that I never discovered, did not speak (to me at least) of her husband, ever. To me she was a rather distant person who lived with my Aunty Mary and Uncle Leslie a half-hour bus ride away in Armley. I would often go to see them and my cousins on Sunday mornings and be home in time for dinner, always the best meal of the week.

Mum would roast a piece of meat, preferably beef, but the first course would always be a large helping of Yorkshire pudding with, without question, onion gravy. Traditionally in Yorkshire it was eaten on its own to fill you up as there might not be much meat to go round. The Cravens kept that tradition going. Even though my daughters were born in the south they had Yorkshire pudding as the first course at home and thought it odd when, elsewhere, roast beef was served with the Yorkshires. But we forsook what in my childhood had been the traditional accompaniment – a bottle of Tizer, or sometimes dandelion and burdock. I recently had a glass of dandelion and burdock for the first time in sixty years and couldn't believe how sweet it was.

My other grandmother, Fanny Noble, was also a widow and she lived in a small cottage not far, as I have already mentioned, from Walter Bennett's butcher's shop in Far Headingley. I was much closer to her than to my other grandmother and sometimes spent the night at her cottage to keep her company, although she preferred to live on her own.

Headingley County Primary School in Bennett Road was just a short walk from home and its teachers opened generations of young minds, including mine and Jean's, to the wonders of learning. I loved every moment of my time in the late-Victorian building, with a large concrete playground at the front and outside loos at the back.

Thanks to our headmaster, Mr Pepper, I made a discovery that would affect the rest of my life – that you can actually make money from writing down words. This was despite his verdict in my school report that: 'John's brain works faster than his hands' – a truth that wasn't overturned until I learned to type.

When I was nine he entered me and some classmates in a national essay competition run by Hovis. I can't remember what the subject was but we didn't have to write very much and though I was not one of the big winners, I did receive a prize of ten shillings (about twenty pounds these days) and a supply of miniature Hovis loaves. My first fee.

In my final year I was appointed infectious disease monitor – quite an honour – and every Thursday afternoon I went into each classroom armed with my register and pencil and noted every pupil who was missing because of a nasty illness, and there were lots of those around.

The only trouble was that I developed eczema on both hands and looked like a walking infectious disease myself, with fingers so swathed in bandages I could hardly write down the absentees. All rather embarrassing, but it got a few muffled laughs from my pals.

I was never top of the class (it's always been my belief that children born in August suffer under our education system because we are always the youngest in our year – and it doesn't help if, like me, you've always thought of yourself as a late developer) but I managed to scrape through the 11-plus and became a grammar-school boy.

Every day I caught the tramcar for a couple of miles (a penny each way) from home to Leeds Modern School. It was a fine red-brick building built in the early 1930s, with large windows and a sports field which it shared with its twin next door, Lawnswood High School for Girls. But boys and girls were strictly forbidden to mix; we were in parallel universes on the same site. A swimming pool linked the two buildings; the girls used it for two and a half days a week and the boys had it for the other half. And no peeking!

The only time I got the cane was after one of the mistresses spotted me one wintery day crossing the line that divided the two schools down the centre of the shared sports field. Woe betide a pupil of either sex who deliberately or accidentally stepped over that demarcation mark, and I had done so at playtime to get a better shot at the girls with my snowballs. Our headmaster, Frank Holland (nicknamed, of course, 'Cheesy', because we thought that was what the Netherlands was famous for) did concede the line was hard to detect in the snow, but Lawnswood was insisting on punishment. 'Cheesy' gave me three whacks instead of the customary six with what, despite the pain, seemed to be a lightish cane and he continued to be one of the most formative figures in my life.

He came to my rescue during one general studies lesson because he must have guessed I was being teased, if not a little bullied, because of my surname. The cowboy movie *High Noon*, starring Gary Cooper, was a big hit and its theme tune, 'Do Not Forsake Me, O My Darlin'' included the phrase 'or die a coward, a craven coward'.

Just about the whole school had seen the movie, so inevitably I began to be branded 'Cowardy Craven'. Ironic, really, as my mother's maiden name was the opposite: Noble. But Mr Holland pointed out to the class that, far from being cowardly, Craven was one of the oldest and most respected of Yorkshire names and a vast area of the county centred around Skipton was known as the Craven District. He thought the name might have come from an ancient word for stony ground and had nothing to do with Gary Cooper and a bunch of cowboys. The name-calling stopped.

There were around twenty classrooms for the 700 boys, linked by brick corridors with oak parquet floors, along which you were forbidden to run or raise your voice. Many years later, after the Modern had been merged with Lawnswood High to form a mixed comprehensive, I returned to present prizes and was amazed to see pupils loudly chasing each other down the corridors at high speed. 'Cheesy' would not have approved.

I have a fond memory of our elderly chemistry master trying to explain some complicated theory by always saying: 'Now watch the blackboard while I go through it.' I don't think he ever understood why generations of boys thought that funny. Sex education consisted of the biology master lowering the blinds in his laboratory, presumably to spare any blushes, and talking about the behaviour of rabbits and waiting until we were married. That was it!

Once I got a detention for daydreaming during a history lesson. I'd been looking out of the window to the road that ran past the far end of the playing field and was thinking: *One day I'll be driven along that road by a chauffeur and I'll give the school a wave.* And one day several decades later I was, and I did.

But I couldn't repeat that some years afterwards because as I passed the school's main gates, to my astonishment the old place had vanished. The great edifice, which, according to the school song, stood 'four-square to all the winds that sweep the northern skies' had been blown away by a demolition gang and its replacement stands on what was once our sports field.

My first encounter with a microphone came when I joined an after-school drama group run by our English teacher. We acted out the adventures of every boy's pilot hero, Biggles, into a massive wire recorder. The gung-ho stories were written by Captain W. E. Johns, who must have really liked his creation because he produced no fewer than ninety-eight Biggles books. I'd read some of them courtesy of Headingley public library and we had great fun turning James Bigglesworth, his friends and his enemies into characters with voices – then hearing our efforts being played back.

We thought, in our innocence, that the recordings sounded rather like a drama on *Children's Hour* on the BBC Home Service, but that was a massive conceit. No way could our stumbling if enthusiastic attempts compare with that wonderful radio programme which was a must-listen for millions of children every schoolday at five o'clock until the emergence of television slowly made it redundant.

Like everyone else, I had my tea listening to the adventures of Norman and Henry Bones, the boy detectives; found similarities between my own classroom experiences and those in 'Jennings at School' (even though his alma mater was much more elite) and let my imagination fly while 'Wandering with Nomad' through the countryside – all the time being guided through that special hour by amiable figures like Uncle Mac. How could I know that children's broadcasting was to exert such a defining influence on my life.

Maybe it was the Hovis prize at primary school or the fact that a much-lauded old boy of Leeds Modern, Guy Schofield, was at the time editor of the *Daily Mail*, but I was beginning to take an interest in the news. It was heightened when, on a grey February afternoon in 1952, Mr Holland called the entire school into Assembly. Something must be amiss because we never had assemblies in the afternoon. He told us that the King had died.

Seven hundred young faces reflected the shock being felt throughout both the nation and the Empire. I don't think we even realized that King George VI had been ill, and now we learned that the young Princess Elizabeth, who was visiting Africa at the time, was to become our queen. At the age of eleven and a half it was the first time a major news event had shaken me and I bought a copy of the *Yorkshire Evening Post* on the way home, something I had never done before.

Then came the build-up to the Coronation, with commemorative coins and mugs and special events at school and Union flags everywhere. On the great day itself, 2 June 1953, street parties were held across the nation, including one in Grimthorpe Street, and they carried on despite the rain. But a large slice of the population also spent a large part of the day doing something spectacularly new – they watched the splendid ceremony unfurling on television – which was quite an achievement because sets were few and far between. Those who did own or rent one opened their doors and let the rest of us into their crowded living rooms. The Coronation was one of the first major events to be broadcast to the masses and we were all keen to catch a glimpse of our new queen.

My family watched with about twenty other people at the home of friends a few streets away, on a set with a tiny screen that had a large magnifying glass clipped to the front. This made the black-and-white picture considerably bigger, but it also emphasized the 405 lines on the screen. It was a magical experience none the less, and we all knew we were witnessing history being made.

The prime minister, Winston Churchill, had objected to cameras in Westminster Abbey because he felt it would diminish the occasion, but thankfully the Queen overruled him. Though he was fully occupied on the day, it's a pity Churchill could not have been with us in that terraced house in faraway Leeds to see the pride and happiness generated by the grainy picture in the corner of the room.

Not long afterwards the Cravens got our own television set, a Pye 14" with a distinctive black facia. Crucially, it was made in the UK – buying a Japanese set would have been the equivalent of betrayal, so far as our family was concerned. Indeed, no Japanese-made product was ever knowingly allowed through our door. Recently, my sister Jean and I went to the National Media Museum in Bradford to record our Coronation memories for a programme I was making about the Queen, and to our delight there on display was a Pye 14" – now a museum piece, like so much from our childhoods.

On top of all the excitement on Coronation Day came news that Edmund Hillary and his Sherpa, Tenzing Norgay, had become the first climbers to conquer Mount Everest, which added to the national sense of well-being on what was a glorious day. It sent me scurrying to our world atlas to find out exactly where the mountain was, and I well remember the headline the following day in the *Daily Express*: 'All This and Everest Too'.

I found myself becoming more and more interested in newspapers and, on a whim, I set about producing a very occasional, very local newspaper of my own which I called the Grimthorpe Street Gazette.

Newspaper is too grand a word – it was two sheets of lined paper from a jotting pad, handwritten by me in pen and pencil with a couple of drawings by my cousin Christine, who lived two doors away. In the Gazette I wrote about events like the big bonfire party on 5 November in our street (it was cobbled so there was no risk of the surface melting).

We kids would make the guy with help from the mums and push it round Headingley in a pram, begging for 'a penny for the guy'. With the money we bought boxes of fireworks and packs of sparklers. Today, you have to be eighteen to buy them and bonfires often involve safety officers and sometimes admission charges. Back then they were literally a free-for-all.

For weeks before the big day anything that would burn was collected – cuttings from trees, old planks, unwanted furniture – and the grown-ups would assemble the bonfire, with a little help and interference from us. In hindsight, it was probably dangerously close to the houses, or perhaps that's just me looking back from these health-and-safety-conscious times.

On the night, chairs were brought out and circled the fire so the oldies could enjoy the fun. By then they had probably forgiven us for the naughty things we had done twenty-four hours earlier, on what we in Yorkshire called Mischievous Night – knocking on doors and running away, tying dustbin lids to door handles, daubing other handles with treacle and lots of other naughty pranks.

As the flames soared the guy toppled and the fireworks started. Rockets were launched with no guarantee they would go straight up into the sky and sometimes they didn't. Catherine wheels were fixed to gateposts and bangers went off everywhere. The people of the street came together in a spirit rarely equalled – and it was sealed with parkin, toffee apples, baked potatoes from the embers and lots of pop. Though the night was high-spirited and packed with potential danger, I don't remember anyone in our neighbourhood ever getting seriously hurt.

But because, for the rest of the year, not much of note happened, and as it could be hard to fill the two pages of the Grimthorpe Street Gazette, sometimes I copied often irrelevant stories from the *Yorkshire Evening Post* to fill the space and I also invented advertisements. Tizer, my favourite drink, got a half a page free, though sometimes it gave way to dandelion and burdock.

Then, with the one and only copy of the Gazette in my hand, I called at every house in the street – maybe thirty of them – and invited people to read it in their doorway for a penny. Almost everyone paid up. I donated the money I made to the RSPCA – so not the beginnings of an avaricious media magnate.

After a while I tired of the Gazette, but not before it had taught me the power and impact of the printed word. One of our neighbours had a rather bad-tempered, snappy dog which she always made sure was safely shut away before she opened the door.

Once, though, it slipped out and tried to attack the

postman. This was a big story for the Grimthorpe Street Gazette. It made headline news and I wrote about it terrorizing the neighbourhood with the postman fearing for his life. Good journalistic overplay and, in my innocence, little did I expect the reaction when – dog nowhere to be seen – the owner in question opened the door, smiled benignly and started to read . . .

The smile soon disappeared. She flew into a rage, threw the Gazette back at me and chased me down the path. If the dog had been loose, I suspect she might have set it on me. It was my first really useful lesson as a would-be journalist: people can be very anti-press unless you write positive things about them. The RSPCA missed out on a penny.

Despite that confrontation, I hadn't lost interest in the news and for my thirteenth birthday I asked for a present that would combine my growing interest in journalism with the fascination I'd discovered for broadcasting during those Biggles recording sessions at school – a microphone. To my joy, my parents bought me the proper job, a Selmer, just like the professional commentators used, with a protected mouthpiece and a wooden handle.

So what else could I do but set up my own news studio – in the kitchen. I got the idea of running the microphone wire through to the living room and plugging it into the wireless. Then, with Mum, Dad and Jean as my first listeners, I'd clear away the tea things from the table, spread the *Evening Post* in front of me and read out the most interesting stories.

I loved the whole concept – selecting the news, changing some of it to fit the spoken word – but little did I realize that a future newscasting career was being born.

Though the Grimthorpe Street Gazette was short-lived, my

interest in newspapers switched from writing them to delivering them, earning a few shillings a week with a morning paper round from Stirk's newsagent on Ash Road – which I noticed on my last visit is now a student letting agency. One of my customers was the Leeds United manager Raich Carter, who lived in a large detached house overlooking Beckett Park. As I was a fan, I'd check the back pages of his papers and if there was good news for Leeds I'd thrust them through into the hall. If it was bad, I'd ease them slightly into the box making them harder to pull out. Just a small gesture of support.

Another source of pocket money was to be found at the Headingley Cricket Ground, just a short walk from home. I'd go there in the late afternoon, when admission was free, and collect the cushions that spectators had left behind after hiring them for the day to protect their backsides from the hard concrete seating. There would be a tip from the ticket office for handing in the cushions and I'd also collect empty beer and pop bottles and get the deposit back from the off-licence. Sometimes I even got to watch half an hour of the cricket, catching glimpses of legends such as Brian Close, Norman Yardley and England's greatest-ever bowler, Fred Trueman, in action.

Using my savings and a little help from Mum and Dad, I bought a portable record player, a Philips DiscJockey. It was to me the ultimate in cool, with a speaker in the lid and a lightweight head that could play both 78 and the new 45 rpm records.

My DiscJockey became the first element in the essential kit I needed to become a fully fledged member of an important new social group invented by advertising copywriters and record pluggers. My 1950s generation were the first teenagers.

Others essentials would later include a Tony Curtis DA haircut (short for duck's arse, because it looked like a duck's tail from the back) that I acquired during the summer holidays. When my dad saw it he was appalled and made me try to comb it out. He didn't mind as much when, a while later, I asked for a four-button-fronted Italian-style suit from Fifty Shilling Tailors.

I was in the early stages of becoming a Mod. Our local cinema, The Lounge, was where I saw my first cinemascope film and everyone in the audience was astonished as the curtain kept going back and the screen, previously a small box shape, got wider and wider. It was also where I had my first date, but as I slowly started to put my arm around her shoulders I saw in horror that sitting just behind us was my rather starchy next-door neighbour, Mrs Wood.

I quickly sat up straight and pretended I didn't know my date, much to her annoyance. She ran straight home and that was that; I never even got the chance to explain to her what had happened. Many years later Noel Edmonds and I were chatting on Saturday morning television about our first girlfriends and I told this story. A few days later I got a letter from her young daughter saying her mum would be grateful if I didn't mention her again. Our other cinema, the Cottage Road, was one of the few I have ever come across with double seats and, when I grew more confident, I found them most comfortable!

Apart from Leeds Modern (so named because it was founded in 1841, whereas its great and distinguished rival Leeds

Grammar School dated back to 1552), the other building that dominated my childhood was South Parade Baptist Church in Headingley. Just round the corner from the cricket ground, it had a large hall used by the youth club and, in the basement, a fully equipped gymnasium where the Cubs, Scouts, Brownies and Girl Guides met.

I'd joined the Cubs when I was eight and soon became great pals with John and David Beal, twins who were the same age as me, and together we became sixers in the Cubs and then patrol leaders in the Scouts. John was in charge of the Bulldogs, David had the Kestrels and my patrol was the Wolves in the 17th North West Leeds.

We loved our Friday night meetings in a gym which had a trapeze, parallel bars and a vaulting horse – so plenty of chances to let loose our energy. Afterwards we went to the nearby fish-and-chip shop on North Lane and each got a bag of chips with scraps for threepence. Then we'd buy a packet of five Woodbines and, coughing fiercely, share them behind the cycle sheds.

As well as bonding in the Cubs and Scouts, our little gang would go to a local stream called Meanwood Beck to catch minnows and sticklebacks and collect frogspawn, or play rounders and tennis in Beckett Park. Sometimes we ventured beyond the confines of Headingley. The beauty of Wharfedale wasn't far away by bike and some weekends were spent camping on top of The Chevin, the steep hill above the little town of Otley in lower Wharfedale, and climbing Almscliffe Crag on the other side of the valley.

Those were the days when parents weren't too worried if you came home with a few cuts and scratches and we were allowed something that is sadly denied to many children today – the freedom of the outdoors. It was on those 'expeditions' that I first discovered the majesty and wonder of the countryside, just down the road from suburban Leeds.

Once, in our mid-teens, we joined some experienced cavers to explore a network of caverns near Kettlewell, climbing down waterfalls and marvelling at how the narrow natural passages we crawled through on our stomachs opened out into cathedral-like caves. One of the team pointed to an extension to the network that had only recently been discovered and we made our way along it, feeling like explorers.

The lights on our helmets picked out all the glistening details until we could go no further. We were several hundred feet underground and in a place that clearly few people had been to before. Excitement swept our cramped bodies until the lamps picked out a sign on the cave wall put there by the local council: 'Please Leave No Litter'. So much for the spirit of adventure!

John and David were proof of how unfair the 11-plus system could be. While I was a moderate achiever at my grammar school the twins, who went to one of the new secondary modern schools, turned out to be far more academic. John got a degree in geology and moved to Australia and David became an accountant.

Together with other members of the youth club, we decided that what South Parade church needed was its own newspaper, and so the Parader was born, with me as the editor. Unlike the hand-written Gazette, it was produced on a very old Remington typewriter using just a couple of fingers. But we got really cutting edge by typing our stories onto an A4 stencil so we could then run off two or three dozen copies on a duplicator. This was big time!

As well as reporting all the run-of-the-mill local church news, we asked members of the congregation to give us their own stories. John and David's dad, who was one of the deacons, recalled driving a fire engine during the war and wrote about a dash over the Pennines to Manchester which ended when the engine in front of his fell at high speed into a bomb crater. Luckily, no one was hurt.

But as well as items like that we decided the Parader could do with a little stardust by way of interviews with well-known names. Our first big scoop was with one of my all-time heroes, John Charles, the legendary centre forward for Leeds United. He was known as the Gentle Giant because he was 6'2" tall and throughout his long career was never sent off or even cautioned – a true sporting gentleman.

He was also a great ambassador for United and one evening he came to talk about his life and career to a large group of teenagers. As he arrived I managed to ask him if, should there be time afterwards, I could have a quick interview. He promised I could, and kept to his word. It's hard to describe just how big a star he was – one of the finest footballers of his era who in the 1953–54 season scored a club record-breaking forty-two goals.

I was a bit tongued-tied in the presence of this sporting god yet he gave me – an unknown kid – enough time to make sure I had some good quotes. And just to make our interview really topical, he was about to become one of the first British players to sign for a top European team, Juventus, where he was idolized.

Another idol who found his way into my 'celebrity column' in the Parader was the Yorkshire-born ballad singer David Whitfield, the first British male recording artist to get a gold disc and conquer America. I first saw him at the Leeds

Empire, a variety theatre that survived through the 1950s despite the growth of television but was demolished in 1961 to make way for a shopping arcade, and three decades later the site became the first Harvey Nichols store outside London.

I loved the whole atmosphere of the Empire – the plush red seats, the brass railings, the orchestra in the pit and, of course, the entertainers, from the dancers who started the show to the comedians who often closed the first half (up-and-coming Des O'Connor was one of them) to the stars at the top of the bill. There was a different show every week and Dad and I, sometimes with Jean, were regulars.

We saw the Beverley Sisters, Harry Secombe, Alma Cogan and the American chart-toppers The Four Aces, as well as three local stars – Frankie Vaughn, Ronnie Hilton and, best of all for me, David Whitfield, with his distinctive tenor voice. I'd bought and played his records on my Philips Disc-Jockey and tried to copy him, and here he was in the flesh.

He descended a twinkling staircase in blue suit and bow tie as the band played the introduction to his signature tune, 'Cara Mia', which had topped the Hit Parade in both the UK and the US. He stepped into the spotlight and the fans went wild – and this was in the days before Elvis and Cliff. Little did we know then that within three or four years their rock and roll would kill off David Whitfield's brand of chart-busting.

Not long after that Leeds Empire performance the Cravens were on a week's family holiday at Bridlington on the Yorkshire coast and David Whitfield was appearing there for one night only at the Spa Theatre. Hours before the show was due to start I hung around the stage door, notebook and pencil in pocket, in the faint hope that I might spot him. After a while, and to my amazement, he stepped out and lit up a cigarette. Hesitantly I approached him and asked if there was any chance of an interview – how about that for bravado from a teenage hack not long out of short trousers?

He asked who I wrote for and I told him the Parader. 'Never heard of it,' he said and I explained it was a small-time church magazine in Leeds. 'Well,' he said, 'I once sang in a church choir in Hull when I was a boy, so fire away.' He told me how he'd sung for his mates when he was in the navy and was discovered on Hughie Greene's *Opportunity Knocks* talent show on the commercial station Radio Luxembourg, which beamed to the UK on 208 metres on the medium wave from the middle of Europe.

Just about every teenager in the country tuned into 'your station of the stars' to hear the latest records being played in a much trendier way than on the dear old *Light Programme* but it also had shows like *Dan Dare* ('pilot of the future'), the Ovaltineys and Leslie Welsh the Memory Man.

My favourite was the Top 20 rundown with Barry Aldiss at 11 p.m. on Sunday, which I listened to in bed. Now that I think about it, the station really missed a trick by starting rather than ending the show with the current No. 1. There was no build-up, no tension, and I was often fast asleep long before it finished.

Anyway, David Whitfield seemed a modest man, even though he was a big star on both sides of the Atlantic. He said he was shocked when the BBC banned his first chart topper 'Answer Me, Oh My Lord', because they considered it blasphemous. He didn't think it was and neither did I (how innocuous it was compared to some of today's lyrics), but he re-recorded it with the words 'Oh My Love' and it went straight to No. 1. My Baptist readers approved of that.

These days such an on-the-spot interview would be impossible because a mountain of agents, managers and public relations people would be in the way. But David Whitfield and I sat on a seaside bench and he gave me as much attention as he would have done a big-name interviewer from a national newspaper. Slowly his star faded, but he was still touring when he died of a brain haemorrhage in Australia in 1980 – and I've always wished I could sing like him.

Alma Cogan was another big star I had in my sights but I almost missed my chance because the stage-door keeper refused to send a message to 'the girl with a laugh in her voice' that I would like to interview her. 'Clear off, you young whipper-snapper,' he said (or words to that effect). Just then a voice behind me asked what was going on and when I turned round there was Alma herself, just arriving at the theatre.

So I was able to put my request to her personally and, pushing past the protesting doorman, she invited me into the auditorium of the Floral Hall in Scarborough, where rehearsals were going on.

Sitting in the stalls, she answered all my questions and asked someone to bring us some tea. Alma, famous for her extravagant dresses and happy songs, could not have been nicer – and impressed me even more by arranging for two free tickets for that night's show.

But back to Leeds Modern and the O level (now GCSE) exams in June 1956. It was a sweltering summer and I'm afraid I spent more time out of doors than inside, revising. And on exam days the sun poured into the main hall and I found it hard to concentrate, especially on our English literature set book, *Northanger Abbey* by Jane Austen.

Could any novel be less appropriate for mid-teen boys than a light comedy about a young woman who adores Gothic novels and spends a social season in Georgian Bath? Much more to our taste was the Shakespeare choice, *Julius Caesar*, with its masculine drama and blood-letting. The exam was a tough one and on the way home there was panic among a group of us on the tram when someone said; 'That was hard, asking us to trace the conversation of the crowd during Mark Antony's oration.'

'It wasn't conversation,' I replied in alarm. 'It was to trace the CONVERSION of the crowd.' Cue mass confusion, but I'm sure I was right, and anyway conversation and conversion resulted in much the same thing – victory for Mark Antony, and luckily I passed.

Not being academically minded, I had already decided I wanted to leave the Modern and earn money in the big world outside – which was fortunate as I managed a total of only five O levels, not enough to get into the sixth form. Career advice was almost non-existent and I had no idea what I wanted to do. For some reason journalism didn't enter my head, even though in those days you could get into newspapers and broadcasting without a degree.

My parents realized that I was not destined for the groves of academe and were happy for me to join the working world – after all, there was no other option. I did toy with the idea of becoming a navigating apprentice with BP tankers, even though my only experience of ships amounted to day trips around Bridlington Bay on board the pleasure boats.

What attracted me was sailing the world and a packing list that included sunglasses and swimming trunks. But then I heard that a navigating officer had lost his job because he failed an eyesight test and I decided that a career which depended on something so unpredictable was not for me. So, in total contrast to the glamour of seafaring, and for want of anything more exciting, I became a commercial apprentice with the Yorkshire Copper Works at Stourton, on the other side of Leeds.

It was a five-year apprenticeship and would involve spending time learning how all aspects of the office departments worked, and some parts of the factory, while also studying at the Leeds College of Commerce for a qualification which some day might lead me to become a company secretary.

The firm made high-quality copper pipes and fittings for the building trade, employed up to 5,000 people on a 120-acre site and quickly opened my eyes to the realities of the grown-up workplace. During the weeks I spent doing 'work study' on the factory floor I quickly came to realize that the slightest upset could lead to industrial unrest.

Hundreds of men worked at machines which robots would operate today – it was mindless, repetitious drudgery so no wonder there was big trouble if the tea trolley was five minutes late. But at least everyone had a job – across the nation there was full employment – and Prime Minister Harold Macmillan told us in 1957 that we had never had it so good. 'Go around the country,' he said in a memorable speech, 'go to the industrial towns, go to the farms, and you will see a state of prosperity such as we have never had in my lifetime – nor indeed ever in the history of this country.' Sixty years after that speech, researchers at Warwick University were to conclude that 1957 was Britain's happiest-ever year.

It was certainly a good one for me, with money in my pocket, lots of friends, my first exotic food (chicken and sweetcorn soup, sweet and sour pork and fried banana for three shillings and sixpence at one of Leeds's first Chinese restaurants), jiving during lunch-breaks from college in the Mecca Locarno Ballroom and just being around when Bill Haley and Elvis Presley burst onto the music scene and shook everything up – especially Elvis with his gyrating hips and first No. 1, what else but 'All Shook Up'? All in all, it felt like a great time to be young.

The six-mile journey by bike from home to Stourton took me through the centre of Leeds, and to cut down on pedalling I treated myself to the latest boost for cyclists, a second-hand Trojan mini-motor. It was a metal box containing a 49cc engine and dinky exhaust pipe and it clipped on to the frame above the back wheel of my bike.

A single control on the handlebar served as both the throttle and the transmission engagement lever that lowered a driving roller, powered by the engine, onto the tyre. It propelled the bike along at a claimed maximum speed of 30 mph, though I never managed anything like that. Disengage the lever and you could pedal the bike as normal.

I had to be careful not to get stuck in the tramcar tracks that still existed in parts of the city but the mini-motor certainly eased my journey – for a while. Sadly after a few months the driving roller began chewing up tyres at an alarming rate (maybe that's why the mini-motor was just a mini-entry in the history of propulsion) and I ditched it in favour of a more traditional motorized two-wheeler, a BSA Bantam 125.

By a piece of good luck I passed my driving test. I had to ride around the main block of Leeds University while the examiner stood with his clipboard at the front of the building. On a road at the rear of the building, well out of his sight, a pedestrian suddenly stepped in front of me and I executed a perfect emergency stop apart, that is, from falling off the bike.

I dusted myself down, wondering whether it had been part of the test and the pedestrian, who was unhurt, had been working with the examiner. If so, I had failed. But it turned out the man was nothing to do with the test because the examiner made no mention of the incident and passed me.

My friend and fellow commercial apprentice Mike Judge had a much bigger and smarter bike, an Ariel 350. By this time the Craven family had moved from Grimthorpe Street to a new semi not far from my old school, and every Saturday morning I would bike over to Mike's home a few miles away and we would work on our machines while listening to *Saturday Club* with Brian Matthew on the Light Programme. During one of these sessions we decided we would take our bikes to the Isle of Man to watch the TT road races.

It was a long way from Leeds to Liverpool, where we caught the ferry, and to my embarrassment Mike, on his powerful bike, had to keep waiting for me to catch up on my little Bantam. But the journey was worth it, the races were exhilarating and afterwards we joined thousands of other bikers riding round the circuit pretending to be TT stars but, in my case at least, without bending over at forty-five degrees round the corners. I wasn't that brave.

Fast forward nearly sixty years and in 2014 I was back astride a BSA Bantam 125 on the Isle of Man, filming for *Countryfile*. I'd mentioned to our team that I'd once ridden round the island and they borrowed the machine from the local motorcycle museum. As it was carefully wheeled from a trailer the years rolled away and those long-forgotten skills came back without even having to think about them. Soon I was whizzing along a cliffside road imagining I could just about make out Mike Judge half a mile ahead.

But to return to the Yorkshire Copper Works, I was starting to realize that a commercial life was not for me. I was far more interested in what was happening in the wider world than in the confines of the factory, which had by then been amalgamated into ICI to form Yorkshire Imperial Metals.

Fortunately my mentor as an apprentice, Frank Clarke, was also the editor of the works magazine and I told him about the Parader. So he allowed me to continue the 'star interviews' I had begun in the church magazine but which would now reach a much wider readership in the *Yorkshire Imperial Mail.*

My first column, in July 1958, featured a fifteen-year-old pop sensation from Scotland called Jackie Dennis who had been on the bill at the Leeds Empire. Our interview took place in his dressing room, carefully monitored by his agent, Eve Taylor. He always wore a kilt on stage ('It's not a gimmick' he protested when I suggested it was) and though he was earning £1,000 a week he told me he only got £2 pocket money and the rest went into a trust.

He was discovered by comedians Mike and Bernie Winters, who contacted Miss Taylor and she fixed up an audition with Decca. His first record, 'La De Dah', reached No. 4 in the Top 20, selling 200,000 copies and for one week he was ahead of Elvis Presley's 'Jailhouse Rock' in the charts. Within three months he had appeared on the top TV music show *Six Five Special*, toured the music halls and appeared in a film. Here's an extract from the interview:

'Would you like to hear my latest record?' he asked, putting it onto the turntable of a portable gramophone he had just bought.

It was a private recording – the commercial disc was not to be cut until three days after my interview – and an arranger was crouched in a corner of the room composing the musical score. It is a silly song called 'The Purple People Eater' and it's all about a man from another planet who invades Earth just to play in a rock 'n' roll band, but it has a catchy tune and plenty of beat.

I'll stick my long neck out and predict it will be his best-selling record yet. [It wasn't . . . so much for my career as a pop pundit.] Before show business overtook him he wanted to be a plumber but, as Jackie says, there are pipes connecting singing and plumbing. He does not smoke – 'He's only fifteen,' Miss Taylor reminded me – and he has no girlfriends apart from Miss Taylor and his mum, who travels with him everywhere.

Frankie Vaughn and Pat Boone are his favourite singers but when I asked if he tried to copy either of them Miss Taylor was most emphatic. 'No of course he doesn't. He doesn't have to.' If rock 'n' roll dies out, as it is sure to do, he is young enough to adjust his talents to another medium. [Did I mean plumbing?] Whether he is just an overnight star who will wither in the heat of big-business show business remains to be seen but he is loaded with talent and deserves all the success which may come to him.

Months after our conversation Jackie went to the United States and was a guest on the top-rated *Perry Como Show*. But his fame was to be short-lived and he faded from the public eye after a couple of years. I'll always be grateful to him, and the redoubtable Miss Taylor, for briefly letting a lad from the Copper Works into their glitzy world.

For my next column in the *Yorkshire Imperial Mail* I met a comedian who remained a much-loved star until he died, Benny Hill. In the summer of 1958 Benny was appearing in a show in Scarborough and I biked over to the coast from Leeds with Mike, having written on company notepaper an official request for an interview. No more hanging round the stage door! I still have a copy of my column and here it is in full:

> He started his career as a comedian in 1941 when, at a very tender age, he toured the working men's clubs billed as 'the boy wonder'. He gathered his material from the repertoire of comedians appearing at the local music halls.
>
> 'I used to listen to the comics at the theatre, copy down their best jokes and use them myself. Club comics today do the same with my jokes,' he told me.
>
> He took two cigars out of a box, offered me one, and lit them both. 'You know,' he said, 'those were the days. After my act at the clubs the secretaries would come up and ask me how I wanted paying. The alternatives were usually half-a-crown or a bottle of pop and a ride home in a taxi. I always took the half-crown and walked home.'
>
> At 18 Benny was called up for National Service and spent three years as a mechanic. For the last two years of his army life he toured with the show Stars in Battledress. 'I gained a great deal of experience from that,' he said. After demobilization he went back to his old job as a club entertainer. He graduated from the ten-shillings-a-night jobs to the three-pounds-a-night clubs.
>
> Then he took a chance and went to try his luck in London where club entertainers are paid the highest fees in the

country. At last came his first appearance on the BBC as a comic on a sound radio series. Since that day the sweet smell of success has surrounded Benny with his own series on TV, his own West End and summer shows, and his own gramophone records. 'I'll do anything for a laugh,' Benny told me, drawing on his cigar. 'I like doing comic sketches with characters that I think up, such as Fred Scuttle. (The funny old man that appears in many of his sketches.)'

Coming into the classification of anything for a laugh are Mr Hill's two gramophone records. 'They were only made for a bit of fun,' he smiled. 'The first one sold twelve discs, including those my mother bought. The second was much better and sold fourteen.'

'What would you be if you weren't a comedian?' I asked, chewing on the end of my cigar. (It's strange how self-assured a cigar makes one feel.) 'Oh, out of work,' was the curt reply. 'I'm lazy. I don't know what else I could do. At the moment I do two shows a night for six nights and on Sundays I have a jolly good rest. I laze around, read the papers and eat. None of these Sunday night packet shows or television appearances for me. I was asked to do a "Sunday Night at the London Palladium" but I refused.'

A question which an interviewer MUST ask a star is 'What are your future plans?' Benny's plans are: a new series for ITV starting shortly and a season in the West End of London which ends in the second week after Christmas. 'After that I don't know what I shall do. Back to the working men's clubs if I'm not careful,' he said nonchalantly.

Mr Hill surprised me by making a revolutionary statement about the variety theatre. 'It's dying at the moment and it will soon be dead. Soon there will be no more variety theatres in the provinces. But packet shows will take the place of the music hall. Every month the big city cinemas will present a variety show for a week that will star many well-known entertainers. One-night-stand shows are very popular at the moment.'

Well, he was right about the future of music hall but wrong about his recording career. Thirteen years later his best-known songs, 'Ernie (The Fastest Milkman in the West)' was the Christmas No. 1 and Benny Hill became a world-wide star. I should have kept that cigar butt as a memento.

3. *The Sunday Break*

When I was barely nineteen years old my career in television seemed to be over – and it had started only a few months earlier after I had written a review in the Parader about a pioneering programme on ITV called *The Sunday Break*.

It's hard to believe in these days of non-stop, round-the-clock television that in the 1950s daytime TV didn't exist and every Sunday evening screens went blank for seventy minutes so that people could go to church – the Evensong close-down. That ended in March 1958 after the moguls of ITV persuaded the Post Office and the Independent Television Authority to let programmes be transmitted during that gap if they were of a largely religious nature and – a major concession by the broadcasters as adverts were still a novelty – free from commercials.

The deal was done and what became known as the God slot was created. One of its mainstays was *The Sunday Break*, which went out live from 6.15 to 7 p.m., three weeks out of four. Concerns had been raised that teenagers would be the hardest to attract to the God slot as we were far more interested in rock 'n' roll and skiffle than in any form of religious broadcasting.

The Sunday Break set out to redress that and as a kind of youth club of the airwaves it had the perfect excuse to include pop stars, special guests and a fair bit of bopping. The whole thing was well meant and rather fun and a perfect subject for an article in the Parader. Sadly, I don't have a copy of my review, but it must have been favourable because, after all, this was the first programme to give young people the chance to get their opinions heard right across the nation.

Hoping for some background on the show, I had written to Arthur Jones, the ABC-TV press officer, and he asked me to meet him in his Leeds office. We had a pleasant chat surrounded by huge pictures of ABC's stars and he gave me lots of info and handouts. Sometime later, and much to my astonishment, I got an invitation to appear on the following Sunday's show to talk about the Parader and give viewers some tips on how to produce their own youth magazines.

Mr Jones drove me to the studios, which were in a converted ABC cinema in the Manchester suburb of Didsbury. I stepped straight from the car into a totally unreal world and was so overwhelmed with excitement that I could hardly take in what was happening. Television was new and magical and just for a few minutes I was going to be a very tiny part of this black-and-white phenomenon.

The whole place seemed a manic whirl of activity with massive overhead lights being adjusted, sets and cameras being moved around and actors, musicians and technicians preparing themselves for that tense moment when the studio went live on air.

What had been the stalls of the cinema was now the main studio and from there Saturday night variety shows and Sunday night dramas were transmitted. The smaller studio, formerly the circle, was home to *The Sunday Break*. After meeting the presenter, Julie Stevens (who was later to appear on *Play School* and *Playaway*), and the rest of the cast, some of them young actors and others TV novices like me, I got a briefing from the show's creator and religious adviser, Penry Jones. He was warm and welcoming and told me to relax and enjoy myself, just as if I was at my local youth club.

Then we did a 'run-through' of the programme and Julie and the producer spelled out the kind of thing I'd be asked about the Parader. I'd also get the chance to join the regulars when they chatted to the guests but I would have to be quick off the mark because everyone wanted to get in at least one question.

When those guests began to appear in the studio, any semblance of an ordinary youth club went out of the window. How about this for a line-up: the Revd David Sheppard, England Test cricketer later to become the Bishop of Liverpool, American actor Sam Wanamaker who even then had dreams of building the Globe Theatre in London, and the 'matchstick men' painter L. S. Lowry, whose work was to become even more famous in the ensuing years.

Before I knew it the red light was on and the show was underway. I did get to talk for a very short time about the Parader and got involved in interviews with Sam Wanamaker, who was starring in *Othello* at Stratford that season, and L. S. Lowry. He told me that one day he missed the train to work, and looking from the platform in Salford at all the people passing by he decided to go home and paint them. That must have been the defining day he became an artist.

The programme seemed to be over just after it started – never had forty-five minutes passed so quickly and before I knew it I was on my way home. But not before I had given my first press interview to the TV editor of the *Yorkshire Evening News*, Ronald Stott, who had been visiting the studios. This is what he wrote:

> John Craven is a bra' laddie as the Scots say. Last Sunday this eighteen-year-old sales apprentice suddenly found himself in front of the TV cameras. I say suddenly because he had only four days' warning that he was to appear. But John didn't panic. He didn't even bat an eyelid.
>
> He remained enviably calm and relaxed . . . and consequently his excursion into television was a complete success. I saw John as he emerged from the comparatively small studio, overheated and overcrowded with the distinguished and the unknown and an overwhelming and fear-inspiring clutter of equipment and technicians.
>
> His face was beaded with perspiration. 'An ordeal?' I asked as he mopped his brow. 'Not really,' said John. 'It was those studio lights that caused this [indicating his perspiration]. Phew! There was some heat in there. I had a little flutter in the tummy earlier in the week but not today. Once we got talking I felt all right and thoroughly enjoyed it in fact.'

And so I did, though I was far from being as calm as Mr Stott made out. Mum, Dad and Jean watched at home – all very different from having to listen to me read the news from the kitchen – and they must have been thrilled with my brief appearance because I got the ultimate Yorkshire accolade: 'you were quite good'. Fortunately, the producers must have liked me, too, because I was invited back and became something of a regular over the next few months.

I travelled to Manchester by train on Saturdays and watched wide-eyed as scene-shifters and floor painters transformed the main studio into the sets for the following night's hour-long live drama in the series Armchair Theatre. In the schedules it followed the top variety show *Sunday Night at the London Palladium*, and could not have been more different.

Armchair Theatre was the umbrella title for a collection of often contemporary one-off plays roughly in the genre 'kitchen sink dramas by angry young men'. Don't forget this was not long after John Osborne had electrified British theatre with his *Look Back in Anger*, and as a result Armchair Theatre got massive audiences. It was to run for nearly twenty years and give a whole generation of actors, writers and directors the chance to shine.

Sometimes I would stay for a while after *The Sunday Break* and watch the final run-throughs. One of the most memorable of all Armchair Theatre dramas was *No Trams to Lime Street* by Alun Owen, starring Billie Whitelaw and Tom Bell. It told of some sailors returning to their native Liverpool and facing social and cultural upheaval. The story goes that the Beatles were so impressed with Owen's depiction of Liverpool in that play that he was hired to write the script for their first film, *A Hard Day's Night*.

During the commercial breaks on Armchair Theatre, make-up artists and dressers fussed around the actors as they dashed to the next part of the set. It was wildly exciting to be there in the flesh, witnessing the contrast between the artificiality of drama inside a television studio and the reality that appeared on the screen, bereft of all the studio trappings. One evening I was so absorbed that I almost missed the last train home, fell asleep in the carriage and woke up in the early hours of the morning in a siding. It was a long walk back to the platform. The end of my rather special 'weekend job' on *The Sunday Break* came rather sooner than I expected with a call from a producer asking if, at very short notice, I was available to go to the Isle of Iona in the west of Scotland to make a film about the monastery there. The style of the show was changing and they would be making more films in the future. Sadly I couldn't go because of work – and that was that.

No more invitations appeared and a friend of mine on the programme told me that it wasn't only that filming on weekdays was a problem for me – I was also getting too old. Too old and I was still in my teens! That was a hard blow to take just as I was getting into my TV stride. But I had been lucky to have had even a brief moment of 'fame', certainly among my friends and family, and it had never been meant to last. I was left with some happy memories, no illusions that I would ever return to a television studio, and just got on with life.

The show itself began to creak and was cancelled in 1965, by which time television was beginning to be more relaxed about what could be shown in the God slot. Coming full circle for me, early Sunday evening is the regular time for *Countryfile*.

The main reason for my interest in Armchair Theatre was that I had become a member of the Copper Works amateur dramatic society. Our theatre was the canteen and our plays were a world away from the realism of Alun Owen – my favourite was the Whitehall farce *Love in a Mist*.

Somehow I also found time to act with the Leeds Arts Centre and the Leeds Children's Theatre and quickly learned that it was not always all right on the night. For the role of the puppet prince in *Niccolo and Nicolette* I had to fly off the stage of the Leeds Civic Theatre. I wore a special belt under my uniform which was attached to a wire that led to a series of pulleys, and at the side of the stage the wire ended in a loop.

The stage manager stood on a chair, put his foot in the loop, jumped down to the floor and the momentum whisked me through the air. One night, though, he missed his footing and I was lifted a couple of feet in the air and then left dangling for what seemed an awfully long time until I dropped down and had to walk off looking as princely as possible. Not easy, but the audience enjoyed it.

So keen was I on acting that I thought of applying to the Royal Academy of Dramatic Art (RADA) in London, a move that seemed all the more attractive because my then girlfriend had just won a place at the London School of Economics. But I decided against acting as a career on what seemed like sound economic grounds: it couldn't guarantee a steady wage whereas the Yorkshire Copper Works could – and eventually, so I hoped, could my emerging love of journalism.

4. Early Reporting

To this day it is a mystery to me why I didn't think about a career in journalism when I left school just before my sixteenth birthday. In the fifties – unlike today – a university degree was not essential if you wanted to be a reporter. Indeed, many renowned journalists, including my distinguished contemporary in the BBC newsroom, John Humphrys, went straight from school to local newspapers and continued their education in 'the University of Life', which arguably can be more useful than any degree.

Looking back now, I had shown interest of a kind in journalism with my efforts on the Grimthorpe Street Gazette and the Parader, but becoming a junior reporter had never entered my mind and the teacher allocated to giving career advice wasn't much help and never asked me about my interests.

If he had, maybe I would have mentioned the fun I'd had writing stories and getting interviews with pop stars and things might have turned out differently. Instead I headed to an apprenticeship at the Yorkshire Copper Works and my professional life as a journalist didn't get going until three years later. Fortunately Frank Clarke, my mentor at the Copper Works, said he would put in a good word for me with Bob Stockton, an acquaintance of his who was the editor of one of Yorkshire's leading weekly newspapers, the *Harrogate Advertiser*. For the first time in my life I learned the importance of 'it's who you know, not what you know'.

Harrogate was fifteen miles from my home in Leeds and, though I had never been there until my job interview, I knew it was a posh place with grand houses, spas and big hotels. On my faithful BSA Bantam I rode there along narrow country roads, past the grounds of a stately home, Harewood House, and then on to the busy Leeds–Harrogate main road, all the time wondering what to expect of both the town and my future career if I got the job.

After all, it was a huge step to take, away from a steady career with definite prospects in a large factory to the uncertainties of journalism in a town I didn't know. During the journey, I thought about an episode that had occurred at the Copper Works a few weeks earlier. As part of my commercial apprenticeship I had, from time to time, to be on duty at the daily morning meeting of the managing director and other senior staff.

With another apprentice I stood at one end of the boardroom's enormous table and handed out appropriate notes and letters and collected instructions to pass down the chain to various departments – we were glorified messenger boys, really. One day, a long and inconclusive discussion was going on around the table about what to do with a large quantity of thin plastic tubing the firm had been experimenting with as a possible replacement for copper tubes.

The MD was annoyed that no solution was forthcoming from his top team so he pointed, Lord Sugar style, at lowly me and said: 'Well, boy, what would you do with all these useless plastic tubes?' Without hesitation, from somewhere in the deep recesses of my brain, I said: 'Turn them into hula hoops, sir.' They were all the craze – I'd even seen girls dancing with them at bop sessions at the Mecca Locarno and tried and failed many times to keep one round my waist for more than a few seconds.

'Brilliant!' said the MD, turning to the others around the table. 'Why had none of us thought of that? With an attitude like his, one day that lad will be running this company.' My chest swelled with pride – singled out for praise by the big boss – and the board may well have considered my suggestion because there was a small bonus in my next pay packet. But the MD's prediction was not to be fulfilled and instead I was on my bike, destined for a different future. Was I doing the right thing?

When I reached The Stray, the 200 acres of broad grassland that wraps around much of Harrogate, I knew I was going to like the place. It seemed to act as a cocoon that protected the old-world atmosphere of the town. I rode along a wide street called The Promenade (which apparently in earlier days fooled some visitors into thinking they would be able to see the sea from it, not realizing the town is sixty miles from the coast) and turned left down steep Montpellier Parade to the fine old building that housed the *Advertiser*.

Its imposing façade quite befitted a respected journal in a very respectable town. The last time I passed by, a few years ago, it was a wine bar, the paper having moved to newer premises. I was interviewed by two people: Frank Clarke's contact Bob Stockton, who turned out to be an urbane and friendly man, and the news editor, Harold Walker, who was tough and quite abrupt. But the meetings must have gone well because I took a drop in salary to £5 a week and joined the small editorial staff of the *Advertiser* and its sister paper, the *Harrogate Herald*.

The *Advertiser*, established in the 1830s and so traditional that it still had only advertisements on its front page, was in sharp contrast to the brasher *Herald*, and Mr Walker told me I would have to adapt my writing style to fit both papers. Though everyone else knew him as Lal, we young reporters always had to address him as Mr Walker, and when he wanted me in his office he would say to his secretary: 'Get the boy.'

A big man confined to a wheelchair after polio robbed him of the use of his legs, Mr Walker was a formidable, uncompromising boss who taught me how to be an accurate and, I hope, truthful journalist. His office was at the bottom of the stairs leading to the reporters' room. He kept the door ever open so that we could not sneak past him and had been known to throw paperweights at anyone whose latest work had displeased him.

He made me quake in my boots and I was not alone – anyone of importance in the town, especially councillors, lived in fear of a verbal lashing in the highly opinionated 'Jottings by Freelance' column he wrote every week in the *Advertiser*. But years later he told the *Radio Times*: 'You get so many, but John was a good one. He had to be trained, got into line, but he had plenty of potential.'

On my first day, Mr Walker was off work and one of the subeditors gave me the task of bringing up to date the obituary of the Dean of Ripon – 'just to see what your style is like'. I later discovered it is common practice to give assignments like that to rookie reporters because no pressure is involved – it could be put in the file and checked at leisure, unless the subject matter suddenly dies!

It took me some time to compile the lengthy obituary to my satisfaction and the reporters' room was empty when

I'd finished. As I had no idea where the obituary file was kept, I put my handiwork into an *Advertiser* envelope, wrote Dean of Ripon on the front so I wouldn't lose it and left it on my desk.

Next day I could not find the envelope anywhere. The other reporters were out covering stories so I could not ask them if they had seen it and, though I frantically searched every inch of the room, it had completely vanished. Then the post-room boy came in and solved the mystery. He said he'd seen a letter on my desk to the Dean of Ripon and, thinking I'd intended to add the address and post it, he had done that for me.

Panic! Just imagine the Dean opening his morning mail and reading his obituary! He would be horrified, surely, and demand my dismissal. I decided to tell no one, and instead made excuses to the subeditor before hastily typing it out all over again and then prayed that the original would, by some miracle (for we were dealing with one of God's representatives here) just disappear.

But it didn't. A few days later Mr Stockton called me into his office and pointed to a letter in front of him. 'You will probably not be surprised, Craven, to learn that it is from the Dean of Ripon,' he said sternly. I went white as my worst fears were being realized: the Dean had surely written a furious letter and my career was going to be over before it had the chance to start.

'You had better read it,' said Mr Stockton, with what I thought was the slightest of smiles – but how could that be at this devastating moment? Then followed words I will never forget. 'Dear Editor', wrote the Dean. 'How considerate of you to allow me to read what you intend to print about me when I join my Maker. I return it now with a few minor corrections.'

The Dean, bless him, was pleased to have been consulted and as a result I kept my job – though Mr Stockton did point out it was not company policy to send obituary notices to prospective recipients and to be more careful next time. My first major crisis was over.

As it turned out, 1959 was a newsy kind of year to enter the world of journalism. It was a time of many memorable events, during which we first heard the following words: M1 motorway, postcodes (as an experiment in Norwich), computers in banks (Barclays to start with), hovercraft (a British invention by Sir Christopher Cockerell, so lots of flag-waving) the Mini (which revolutionized motoring, especially for the young) and duty-free airport shopping.

Margaret Thatcher entered Parliament, Ronnie Scott opened his jazz club, racing driver Mike Hawthorn was killed, the painter Stanley Spencer died and Colin MacInness published *Absolute Beginners*, a disturbing novel about young Londoners fearing for the future amid the violence and hatred of the Notting Hill race riots – a book which had a profound effect on many teenagers, including me.

Meanwhile, in leafy Harrogate, a world away from all the urban tensions in the capital, I was learning how to transform small local issues into page leads and building up contacts in the community who could give me those very stories. Daily life included regular calls to the police station, the hospital and the fire station, sitting through endless, often uninteresting cases – story-wise, at least, if not for those involved – in the magistrates' court and collecting the names of mourners at funerals.

'Every name means another reader,' Mr Walker used to say. So, to keep him happy (but without his knowledge) we would sometimes make up extra names if not many people turned up to say a final goodbye to someone who seemed to us to deserve a bigger send-off. Only much later did I consider the implications of this – how relatives might have puzzled about those invented names, wondering who on earth they were – secret figures, perhaps, from the deceased's past? Had our good-intentioned motives created family mysteries?

Slowly I began to realize what a strange trade I was in, attempting to remain even-handed and considerate while reporting 'without fear or favour' the stories that make up the fabric of a town like Harrogate. Sometimes I would be there, notebook in hand, reporting a happy event like a golden wedding and hours later delving into a great personal tragedy and hoping to get a quote. As an old hack once told me: 'It's a bit like walking down a street and through some doorways you throw a bag of gold and into others a bag of rocks.'

With some of the people we wrote about, the last thing they wanted was to appear in print, while others were totally delighted to see themselves in our pages. Working with me on the *Advertiser/Herald* were two other young reporters: Peter Rose, who was local and had been in the job for a couple of years, so knew the ropes, and Sue Francis from Leeds, who, like me, was just starting out.

We all got on really well, in the office and socially – so much so that Peter and Sue later married, while Sue and I acted together in local amateur productions. Later they moved to London, where Peter became a political correspondent and Sue was the head of media for the Arts Council, and they stayed good friends of mine through the years.

Sue got her pilot's licence in her sixties but I have yet to take up her offer of a flight round the Home Counties. I haven't been very good in light aircraft ever since one caught fire as we flew across Lake Victoria while I was filming in Africa. Peter later left journalism for academia and wrote a definitive book on the Northern Ireland Troubles. I had the honour to write his obituary for the *Yorkshire Post* when he died in 2017.

As part of the learning curve I was, like all young journalists, drilled on the importance of correct spelling and grammar. One colon in the wrong place can totally change the meaning of a headline, sometimes disastrously so. A fine example was the headline in one Yorkshire paper when King George VI was lying in state. It should have read: 'King: Moving Scenes'. Instead it said: 'King Moving: Scenes'.

Puns, said Mr Walker, should be avoided (someone tell that to the current subeditors on the *Sun*) but sometimes double-entendres can catch you out when you miss the double meaning. One of our team once wrote the headline 'Councillors change their minds on public lavatories'.

After a few months on the paper, commuting every day, I made a big decision and upgraded my personal transport from two wheels to three. My faithful BSA Bantam 125 was sold and, for £150, I bought a nearly new bubble car. At the time, they were all the rage and mine was a Heinkel, made from the cockpit mouldings of the eponymous German warplane. It had a single door, left-hand drive, three wheels, two seats, a sun roof, a 198cc four-stroke scooter engine capable of, if you really put your foot down, 50 mph maximum, and I loved it.

Because it had belonged to the *Advertiser* and been used by one of the sales team until he switched to a proper car it was in good condition and I knew where to complain if anything went wrong. With the reverse gear blocked off I could drive it on my motorbike licence and on my first day behind the wheel I breezed into a car park and stopped too close to the wall. The front-opening door couldn't move forward enough to let me out and I was trapped. I had to make an undignified exit through the small fabric sun roof watched by a small group of amused onlookers. I never parked too close to anything again.

That door actually proved a godsend to my grandma Craven. Because she suffered terribly from arthritis it was too painful for her to get into a normal car but my Heinkel posed no problem for her elderly body. We just helped her gently backwards though the front door space, eased her into the seat and off we went for a nice little run. She loved the fact that the wraparound windows gave her a wonderful view of the countryside – something she hadn't seen for a long time.

But one winter's day, while on my way to work, the Heinkel and I ended up quite literally as part of the scenery. The roads were covered in snow that had fallen upon ice and, though I was driving carefully, the lightweight vehicle suddenly skidded to the left, went across a verge, and rolled over, rather like a giant snowball. The large bulbous back window blew out and once again I crawled out of my little car, this time through the space where the window had been.

People waiting at a nearby bus stop saw the whole spectacle and two or three came over to make sure I was all right. The Heinkel was in about four inches of snow but together we managed to right it and push it back a few yards on to the road. We were all out of breath – the effort certainly warmed us up – and apart from the missing window the Heinkel was undamaged.

So I put the window into the back and, more in hope than expectation, turned the ignition key. Amazingly, it started first time, prompting a little round of applause from my rescuers, and I resumed my journey to Harrogate even more gingerly. Mr Walker accepted my apology for being late to work and suggested, ever the news editor, that it might make a paragraph in the winter accidents round-up. I sold the Heinkel after about a year when it started to use more oil than petrol in its little engine and splashed out on one of the very first Morris Minis. It was basic and beautiful and cost around £500. After proudly picking it up all shiny and new from the dealers I took my mum and dad for a spin around Wharfedale – and it broke down.

Being a generally respectable and law-abiding place, Harrogate and the surrounding areas of Ripon, Knaresborough and Pateley Bridge rarely had a visit from members of the national press. But when a big story broke they rode into town from their bases in Leeds and Manchester like a posse of avenging angels (or demons – take your pick).

The tactics used by reporters from 'the nationals' were much the same across the nation; they would glean basic facts from local reporters like me then set about getting their own scoops, doing their utmost to outfox and outbid their rivals. Some of their methods were devious to say the least. I heard of one Fleet Street crime reporter who carried a vicar's dog collar in his briefcase. When he arrived at the home of a family bereaved in some awful way, he would clip on the collar and ask if he could be of help and support. That way he got an inside story without ever revealing his true purpose or identity.

From my own experience of dealing with them, I found

the big-time reporters an interesting, eclectic bunch, some of whom looked and behaved like rogues while others looked more like bank managers or businessmen. I once had a glass of champagne (he was paying!) with one of the latter. He was among the greatest-ever feature writers and a hero of mine, Vincent Mulchrone of the *Daily Mail*.

I had read many of his articles and wished I could write like him – and now I was with him for a day as a kind of junior assistant. A big man with friendly eyes and a mastery of the English language, he will always be remembered by people like me for his riveting opening sentences. In 1965, when thousands of ordinary folk were paying tribute to Sir Winston Churchill by filing past his coffin in Westminster Hall, Mulchrone wrote: 'Two rivers flowed through London last night and one was of people.' Brilliant, tear-jerking stuff.

Our meeting had happened a few years earlier in a Harrogate bar. While we enjoyed his champagne (I was told he always drank it for breakfast) I briefed him on the International Toy Fair which was held every year in several of the town's large yet genteel hotels to showcase the must-have gifts that Santa would need to have in his sack the following Christmas. It was a big occasion for the town and an early step in its transition from up-market spa resort to major conference and events destination.

Mr Mulchrone and I went together to the fair and I left him to collect material for his piece, in which he cleverly merged the latest toys on display with Harrogate's reputation as a quiet idyll for well-off retirees, many of them ex-military, who would snooze in comfortable hotel armchairs following afternoon tea.

I can't remember his opening words (I did keep a copy of

his piece for years as an example of how to write a feature article but somehow, like so many newspaper gems, it ended up in a bin – such is the ephemeral nature of journalism), but they were something like this: 'Perhaps it was the invasion of strange flying objects called Frisbees that finally forced the General into retreat.'

Mr Mulchrone had seen all that I had seen at the fair, but with his own imaginative and cleverly written interpretation of the event had taken it to a level I could only dream of. For me it was a masterclass and afterwards I did try to raise my game thanks to a Fleet Street legend who died from leukaemia at the age of fifty-four.

After cutting my journalistic teeth on the *Harrogate Advertiser*, and learning a great deal about the relationship between people and the press, between readers and writers, I left the paper on good terms and spent a few months as a freelance in the town. The huge difference, I soon discovered, was that instead of just typing out your story and handing it in for approval you had to telephone it through to seven or eight different newspapers.

No mobile phones, not even fax machines in those days, so first you had to find a telephone box that worked and then defend yourself from others who wanted to use it for perhaps half an hour. And it was so boring, repeating the same words time and time again.

Men and women – often, it must be said, rather grumpy men – sat with headphones at typewriters in the copy room and took down your dictation and the story was then passed to the newsroom. You had to be precise. If you were quoting someone who, say, wasn't being co-operative, you would say: 'Quote No comment comma unquote he said point [meaning full stop].' It was all very laborious, frustrating and time-consuming, especially if a deadline was looming.

You had to be clear in your delivery otherwise the copy-taker would complain and get edgy. I must have been gabbling one day when a story of mine appeared in one paper saying an eminent scientist had been talking about Youth in Asia. It should have been euthanasia.

My next move was to the *Yorkshire Post*, which thinks of itself to this day as Yorkshire's national newspaper, and I didn't have to move from Harrogate because I became part of its three-strong staff covering a large area of the county from our base in the town. I still had to phone in my stories (emails and texts make life so much easier these days) and on one occasion the copy-taker really gave the game away.

I was covering the annual dinner of the Yorkshire Licensed Victuallers Association at the Old Swan Hotel in Harrogate – my favourite of all the town's large hotels and famous for being where Agatha Christie suddenly reappeared after vanishing for eleven days at the height of her fame in 1926. A nationwide hunt had failed to find her but she had been staying quietly at the hotel under the assumed name of her husband's mistress. It could have been the plot for one of her novels if it hadn't been real.

The Victuallers dinner was, as you might expect from Yorkshire publicans, off-licence shopkeepers and their other halves, a rather loud and jolly affair. The paper took special interest in the event because there was an advertising supplement involved and we dreamt up a story about the hotel's two youngest employees, at chambermaid and a trainee chef, being the guests of honour for the night – the 'Cinderella and Prince Charming'.

Photos were taken, the story was written and then dinner began. I was not much of a drinker, but the flowing wines and spirits got the better of me and I was just about to slide under the table when I realized through an alcoholic haze that I hadn't phoned through my story. Somehow I found my way to a public telephone in the hotel and dialled copy.

The 'seen-and-heard-it-all-before' copy-taker quickly realized the state I was in and said in a weary yet fatherly way: 'Just try to give us the facts, John, and we'll do the rest.' So I stumbled through the details of Cinderella and Prince Charming etc. and then slowly crumpled to the floor.

I don't know how I got back to my digs but next morning I woke in a cold sweat. What had happened last night? Did the paper get the story? Just how much trouble was I in? Racing to the front door, I pulled the paper from the letterbox with trembling hands. To my huge relief there was the story and the photograph, perfect in every detail – except for one. The copy-taker and the subeditor had a joke at my expense. The by-line read: 'Yorkshire Pist Reporter'. How true!

Every now and then, even to this day, the *Yorkshire Post* holds a literary luncheon which is quite literally a moveable feast that ambles from town to town around the county. It has always attracted leading authors keen to sell their latest books and when the luncheon came to Harrogate during my spell on the paper my task (I'd been forgiven for the Licenced Victuallers incident) was to keep one of the star guests sober – and I failed miserably.

The writer in question was Brendan Behan, one of Ireland's greatest poets, playwrights and drunks. The literary world loved him for plays like *The Quare Fellow* and his autobiography, *Borstal Boy*, but the public knew him better for his dishevelled, often tipsy appearances on television screens and in newspapers from Dublin to New York.

I never did understand why this brilliant Marxist republican rebel was invited into that bosom of the Establishment, but it was a scenario doomed from the start. After all, he once said: 'One drink is too many for me and a thousand is not enough. I am a drinker with writing problems.' He was given a suite in the hotel where the event was held and my orders were to be his unofficial minder and stop him, from breakfast onwards, getting close to any Guinness or whiskey – or, indeed, alcohol of any kind. It all started very well. When I arrived he was full of Irish charm and seemingly sober.

We had a long and rambling chat – the nervous young reporter somewhat out of his depth and the famous writer who, as the world knew, was being slowly destroyed by his beloved booze. I told him I had to write a story for the *Yorkshire Evening Post* and could he tell me something about himself that had a local connection – always a necessary question on regional newspapers. 'Indeed I can and it's not off the record – you can publish it,' said Mr Behan, edging towards me in a confidential manner. 'When I was a young man I was in the IRA and I came over from Dublin and tried to blow up power pylons in Yorkshire.'

What an exclusive – Behan the bomber! In *Borstal Boy* he'd written of his idea to target Liverpool docks with explosives but this Yorkshire angle was new to me. I wrote up the story and rushed down to the lobby to phone it through. When I returned to his room he answered the door swaying a little and that rich Irish brogue was definitely slurring.

Though I'd been told no alcohol had been supplied to his room, he had somehow acquired a bottle. He smiled as I tried to tidy him up a little – several hundred guests were expectantly awaiting this thirty-something *enfant terrible* – and down we went to the reception. Once he started mingling there was no holding back. No passing waiter was safe from his outstretched hand and by the time he joined the top table he was fairly blotto.

Other speakers included the television presenter Kenneth Allsop, whom I was to work with years later on the BBC's first environmental programme, *Down to Earth*. The audience loved him – he was witty, urbane and a master storyteller. But when it was Mr Behan's turn you could sense the apprehension in the room; his shambolic behaviour during the meal had been pretty obvious. He rose unsteadily to his feet, gave a short and not too coherent speech and sang an Irish protest song. That was it! Regular luncheon-goers had never seen anything like it – and never would again, especially as he tried to chase a rather distinguished lady around the table.

I saw him later as he was stumbling from the gents toilet. Someone had shown him my 'Brendan the Bomber' story across the front page of the *Evening Post*. He beamed and embraced me and said we must talk again later but we never did. As a reporter, I was grateful to him for giving me a scoop which certainly got the afternoon tea drinkers talking at the Old Swan and stopped my bosses from asking too many questions about my failed mission. I realized that in our few hours together I'd developed a concerned affection for this gregarious, rebellious man whose love for the bottle was both his inspiration and his downfall. Four years later it killed him.

For much of my time on the *Post* I lived during the week at a small hotel called the Burlington, just across the street from my office. I did a deal with the owners to have free board and lodging in return for manning the reception on the evenings that I was not on call to the paper and also working as a breakfast waiter.

It was an eye-opener in many ways. I spent much of one night doing what I could to comfort a man and his young daughter. The man's wife and son had been killed in a car crash on the A1 on their way home to Scotland. They were brought to the hotel by the police as they knew no one in Harrogate and I stayed with them until a relative came to collect them the next day. The hotel owner didn't charge for their stay and for me it was a salutary lesson in the awful reality behind the stories I had so often written about road accidents.

One day when I was on breakfast duty, I went to a table where one of the late arrivals was seated, slightly hidden by a pillar, and said, in a bright waiter-like manner, 'Will it be tea or coffee this morning?' To which the guest replied: 'What the hell are you doing here, Craven? Why aren't you in the office?' It was none other than Desmond Pratt, distinguished theatre critic and a deputy editor of the *Yorkshire Post*.

The night before, I hadn't been on duty when he checked in after reviewing the latest production of the White Rose Players, the repertory company based at the Harrogate Opera House. Desmond Pratt was a much feared, much admired critic and one of the top names on the paper. The others at the time were the industrial correspondent Bernard Ingham, he of the enormous eyebrows who went on to achieve fame as Margaret Thatcher's press officer, and Jean Rook the women's editor, who later became known as the First Lady of Fleet Street (and, less kindly, as one of *Private Eye*'s models for Glenda Slagg).

Desmond Pratt got a review himself in Paul Allen's biography of the Yorkshire actor and playwright Alan Ayckbourn, *Grinning at the Edge*. He said that Pratt's status

> meant that he could get away with arriving late and leaving early; and sometimes he did. He checked out that the actual performance had taken place and then filed copy written beforehand about the play itself, based on a close scrutiny of the script, a memory of other productions and a very fair idea of what the actors you saw on a regular basis were going to be like. For anyone in the theatre who believed he or she was doing something really special just this once, such a procedure could be heart-breaking.
>
> Yet there were plenty of other times when the critic would reveal himself to be anyone's equal as a perceptive and original critic catching the quality of a performance with a skill that was hard to match whether he was praising it or identifying a cardinal problem. His support for Alan as an actor and playwright came early, when it was most useful, and Alan is grateful for it, whatever his general thoughts about critics.

And here was this doyen of theatre criticism glaring at me, in the dim light of the Burlington's breakfast room, as though I was putting on the worst performance he had ever seen. I mumbled some excuse about standing in for a friend, took his order and made a quick exit stage left. His parting words were: 'If I am ever here again, make sure you are not.' I did not do an encore; the manager delivered his order for me and fortunately the great man took the matter no further.

Though freelancing had not really appealed to me, the

offer of slightly more pay and a change of scenery led me to change my mind and join a news agency in Bradford. Most of my day was spent reporting for the now-defunct *Yorkshire Evening News*, starting each day with a check on the magistrates court.

Every Monday morning the drunk and disorderlies from the weekend would be up in front of the bench. Most of them were 'gentlemen of the road' with no fixed abode and the normal routine was for them to be found guilty and given the bus fare to Leeds. Just as in the old Western movies it was: 'Get out of town by sunset.'

One Monday morning an Irishman appeared before the bench in bedraggled clothing and looking much the worse for wear. To everyone's surprise he pleaded not guilty, words rarely heard in such cases. The magistrates' clerk, a somewhat Pickwickian character, asked him if he had anyone to represent him. 'No,' bellowed the Irishman, 'God will be my defender. God Almighty in his heaven will defend me.' To which the clerk replied: 'That's all very well but don't you have someone a little more local?' The tramp was found guilty and got his bus fare to Leeds.

At the other end of the social scale, when the Queen Mother visited the area one evening, one local photographer was given a special assignment. He was told to get pictures of every stage of her visit. And it didn't start well. As Her Majesty stepped from her car he was in a prime position to get his photo – but the flash didn't go off.

He had a very expressive face which must have shown his panic because the Queen Mother noticed and said: 'Shall I wait?' He nodded in gratitude and said: 'If you wouldn't mind, Ma'am.' So, she did, the flash worked and he got a lovely, smiling picture. He went on to be a top cameraman – and one who would never take a bad photograph of the Queen Mother.

Perhaps even bigger than her in the fame league at that time, around 1963, were the Beatles, and as part of a headline-grabbing, scream-drowned tour of the provinces they came to Bradford for one night only. I met them briefly before the show and a 'photo opportunity' was arranged; George would slip out to buy some shirts from a nearby store. Such was the level of Bradford Beatlemania it had to be a carefully controlled mission and even so he narrowly avoided being caught in an avalanche of fans.

What happened that evening was something the city had never seen before. The Gaumont Cinema was jam-packed with teenage girls and they never stopped screaming. Many were in hysterics long before the Fab Four arrived on stage. The warm-up comedian gave up on any attempt to tell jokes and instead endlessly shouted out their names. The noise was above deafening – the decibel level went off the scale.

When the Beatles did appear they sang all their hits but no one could hear them so to me it all seemed rather pointless – the girls had paid good money to be in the presence of their idols yet all those No. 1 hits were lost amid the screams. Towards the end of the show the Fab Four shouted: 'We'll be back in a minute!' But they wouldn't be and by the time the fans realized the gig was over the boys were on their way to their home town of Liverpool, the next stop on the tour.

The aftermath was like a battlefield. Girls sobbing, some lying in the aisles or spread across the seats, others comforting friends, St John Ambulance volunteers treating those who had passed out or were inconsolable. This was

Beatlemania – and it took my eardrums quite a while to recover.

But for our agency there was a bonus. A litter of tiny puppies had been found abandoned – no big deal as sadly it happens all the time – but we picked out the four males, photographed them so they looked at their most appealing, and named them John, Paul, George and Ringo. The picture appeared in just about every paper and the whole litter was quickly adopted. It's that kind of quirky, topical, opportunist twist on a big theme that helps local news agencies to survive.

Bradford was a fascinating, cosmopolitan city of many contrasts: the woollen industry which had been its mainstay was in decline, there was a growing immigrant population from Asia which seemed to be settling in rather well, and a worrying drug problem among teenagers which one national magazine asked me to investigate:

> The long-haired boy weaved his way through the twisting teenagers to a corner of the club where a group of youths were talking. He sidled up to one and whispered: 'Got any?' His friend gave a brief nod, looked quickly around the club and slipped him a couple of purple hearts.
>
> Where did this little scene take place? Soho, where Scotland Yard drug squads have been raiding teenage haunts? No – it happened in the cellar below a Methodist mission chapel in Bradford. A third of the teenagers who go there are drug addicts and many of the others are misfits. They go there because no one tells them to wash or have their hair cut.
>
> Known traffickers are banned in Dave's Cave, which was set up by two Methodist ministers as a refuge for teenage drug users, especially those trying desperately to kick the

habit, and it even has its own band, the Junkies. One seventeen-year-old told me: 'We try to stop drugs being used in the Cave. The great problem here is – can those of us who don't take drugs change those who do, or will they change us? We outnumber them, so I think we stand a better chance.'

I've often wondered what happened to Dave's Cave and the sixty or so teenagers who gathered there every night. One of the pitfalls of journalism is the constant need to move on to the next story without much compunction to follow up on previous ones. That is especially true if the journalist in question moves to a new location, and after a year or so in Bradford I was developing itchy feet.

When the offer came of a new job I uprooted myself from the West Riding of Yorkshire, where I had lived all my life, and moved 120 miles north to Tyneside. And though I didn't know it at the time, it was there that my broadcasting career, which had come to a sudden stop when I was no longer needed by *The Sunday Break*, would start all over again.

5. Tyneside

When you are a freelance reporter you never know what the next phone call will bring and from which newspaper. It might be the *Daily Mirror* wanting more background on a juicy revelation or the *Guardian* enquiring about development plans for a new town. No matter what, you try to provide the information.

Shortly after I started work on Tyneside one of the tabloids asked me to try to contact, of all people, the Prime Minister, Sir Alec Douglas Home at his country seat, The Hirsel, at Coldstream on the Scottish Borders. Not about any vital issue of state – instead, the paper had a great photo of him in action at a local fête. Sir Alec had knocked down a pile of cans with a ball (he had been a first-rate cricketer in his youth) and won a coconut.

The paper wanted a quote, so I managed to find the phone number and fully expected a member of staff to answer my call. 'Could I please speak to the prime minister?' I said, in as confident a tone as I could muster. It worked because the voice at the other end said: 'Speaking.' You could have knocked me down with a feather – the Prime Minister of Great Britain and Northern Ireland answering his own phone!

He was, no doubt, expecting a far more important call but we both got over our shock and when he realized it was an innocent enquiry about his throwing skills he gave me a nice little quote about how lucky he had been.

I had moved to the north-east to work for a news agency based on the coast at Whitley Bay, and one winter's day I was sitting in my car on the seafront eating my sandwich lunch, watching the grey waves pounding, when a strange thing happened. An elderly woman strolling along the sands sheltered by an umbrella suddenly turned sharp right and started walking slowly into the shallows.

For a moment I was too stunned to do anything – could this really be happening? Then I came to my senses, leapt out of my car and raced across the beach towards her. Another man sheltering from the rain also witnessed the drama and joined me as, fully clothed, we waded in and I was almost knee-deep when we reached the woman.

Somehow I expected her to be grateful to her rescuers, but she wasn't. She struggled to free herself, hitting out with her umbrella and swearing as we dragged her back to safety through the bitterly cold water. It was all so unreal. She was still trying to escape back into the sea when other help arrived and eventually she was led away to an ambulance that had responded to a 999 call.

Still in a state of shock we rescuers shook hands and went our separate ways to get into dry clothes and carry on with the day and I later discovered that the woman had absconded from a psychiatric unit.

One of the regular customers for news stories was the BBC, which covered Northumberland, Durham and what was then Cumberland and Westmorland from its radio and television studios in Newcastle upon Tyne. I phoned across many stories to its copy-takers and shared digs in Whitley Bay with one of its senior journalists, Michael Fitzgerald.

I told him about my brief experience in broadcasting with *The Sunday Break* and said I would like to get involved with radio and television news. So Mike arranged for me to have a meeting with his BBC bosses and as a result I was given a 'trial' as a newsreader, on the lunchtime regional radio bulletin on the Home Service. It was a baptism by fire, and though I was shaking with nerves I managed to get through the five minutes with just a couple of small trips. But I did commit the cardinal sin of still speaking when the 'pips', the Greenwich time signal, sounded for one o'clock.

No one seemed to mind though, and I had passed this live audition – those childhood moments of reading from the *Evening Post* at the kitchen table had become a reality, and I loved it. Luckily, I was invited back to read the news a couple of times a week, and though I continued to work for the freelance agency, I also did shifts writing bulletins on the BBC news desk. This new aspect of my career suited me perfectly – the thrill of the studio, the stress of the deadline, the challenge of writing precisely to the second, of getting pronunciation right (something I had never before had to consider). I had seen the light – and it was a green light! I wanted to be a full-time broadcaster.

And I could not have been in a better place at a better time. BBC North at its city-centre studios at 54 New Bridge Street – what had been a nineteenth-century maternity hospital with a modern extension at the back – was going through something of a golden era. A great team of journalists and presenters was reflecting life in a region which, after years in the doldrums of decline and unemployment, was starting to have faith in itself again.

In 1965 I landed a staff job as what was quaintly termed a news assistant – what any other news organization would call a junior subeditor/reporter – and I still occasionally read the bulletins on radio and television.

One thing you quickly learn as a newsreader is to have your wits about you when you're on air, or you'll easily be caught out. The odd spelling mistake can slip through, like on the day when I was handed a newsflash just before transmission which read: 'An RAF rescue helicopter is searching the Solway Firth for two missing wildflowers.' Luckily I twigged that it should have been: 'wildfowlers'.

As well as bulletins for the whole region we also provided a news service on new-fangled VHF (later known as FM) for Cumberland and Westmorland (now Cumbria), and our little joke was that the area had far more sheep than listeners, especially as hardly anyone had VHF radios. In that part of the world they really know how to celebrate New Year and when I was reading the VHF bulletin early one New Year's Day I asked myself out loud, just before the time signal: 'Have you ever had the feeling you've been talking to yourself?' No sooner had I got back to the newsroom than the phone rang and it was a farmer's wife from the Cumbrian fells saying: 'Don't worry, John, we were listening.' That's loyalty for you.

From time to time I also presented the early evening radio magazine *Voice of the North*. It was a show that reflected the mood of the region both socially and politically in no uncertain terms and it pioneered vox pop – from the Latin *vox populi*: voice of the people. Reporters went outside the studios with a very heavy portable recorder, laughingly called the Midget, and searched for the often outspoken comments of people in the streets on current issues – quite a breakthrough for radio back then.

In 1967 I was promoted to senior news assistant and joined the BBC's permanent staff, which, in those days, really could mean a job for life. The old joke was that you would only get the sack if you were caught misbehaving with the Director General's daughter. That golden age for BBC Newcastle was led by the late Mike Neville, one of the best television presenters ever and anchor man of *Look North*. Broad and blond-haired with a ready twinkle in his eye, Mike had the magic touch – gravitas when needed but a warm and friendly style and a brilliantly quick Geordie wit.

The audience could identify with him and he was their nightly friend for thirty-two years. Though the show ran smoothly nearly all the time, Mike was the man who could cope with any crisis and was in his element keeping viewers informed about what was, or was not, happening. One night when I was reading the news smoke started to waft from inside the elderly camera I was talking to. 'Hold on a minute,' said Mike. 'Could the director swing my camera to take a shot of John's because he's about to make news himself!' The fire was quickly put out and the show carried on. It had been a funny yet dramatic moment and Mike's quick response was a classic example of his easy professionalism.

When the telecine machine that transmitted our film reports broke down, which was not uncommon, he would call out to the engineer in charge of it: 'Give it a kick, Flash!' When the communications line went down from the BBC in London to *Look North* and we couldn't show a crucial live interview with the Director General about broadcasting policies for the 1970s, Mike and another outstanding member of the team, George House, filled in for almost the whole programme.

They chatted away about this and that and occasionally checked to see if the line was back again, which it wasn't. Viewers probably found it much more entertaining than a long-winded explanation of the BBC's future plans for its English regions.

Another time Mike had to cover quickly was when we were leading the programme on a demand from some Scottish campaigners that the border with England should be moved further south towards Newcastle. Due to a misunderstanding a rather militant member of the campaign arrived at the studios a couple of hours too early and, as we were all busy, we left him alone in our reception room with a cup of tea. Come transmission time he was really fed up. After a long introduction he was asked where he wanted the new border to be. Pointing at a map he said: 'Between there and there, and good night ter yee,' and walked out. Cue Mike to 'fill' for five minutes.

I learned so much from watching him and George and newsreader Tom Kilgour, a gentleman broadcaster and one of the nicest, kindest people I have ever met, just doing their jobs. Another member of our reporting team was a young man who joined us from Border TV in Carlisle and I could tell immediately that he would go far – he was a good reporter, handsome and ambitious. His name was Michael Rodd, and he went on to present *Screen Test* and *Tomorrow's World*.

Mike Neville could have become a top network presenter, and indeed for a short while hosted *Nationwide*, the hugely successful current affairs show from London that followed the regional programmes on BBC One. They asked him to stay but he didn't want to leave his beloved north-east, and told me once that he liked to spend his holidays 'outside the recognition zone'.

Talking of *Nationwide*, I took part in a trial run for the show and all I had to do was read one line from a famous Shakespearian speech. I sat in the studio and waited for my cue, as did other reporters in every regional studio around the country – and they all had lines from the same speech. The point of the exercise was to see how quickly the vision mixer in London and the studio crews could switch from region to region as we all took it in turns to read our line in the right order. Never before had the Bard's mastery with words been put to such a practical use and it obviously worked because when *Nationwide* hit the screens one of its trademarks was the rapid switching between regions during its nightly round-ups.

I made some great friends in Newcastle and one of the best was a young film editor called David Pritchard, who had moved north from Hampshire and, like me, fallen in love with Geordieland. Just as I dreamt of being a full-time television reporter, David wanted to be a TV director/producer. And he did make it big-time, running some of Britain's top cookery shows and discovering Keith Floyd and Rick Stein along the way.

But back then we conspired to make a short feature film together in our spare time, hoping that it might be good enough to be shown on *Look North* and kick-start our ambitions. David would direct and edit it and I would write the script and be the presenter. First we needed a good story that would appeal to our news editor Terry Dobson, a brilliant but demanding journalist who went on to create *Pebble Mill at One*.

Fortunately, Terry liked us both, so we reckoned we might be in with a chance of making our debut film if we could find the right subject – and it came along in the unlikely form of the *Lizzie and Annie*, a grubby little cargo ship built at South Shields on the Tyne in 1877 and now awaiting the breakers' hammers. The Tyne was the birthplace of many famous vessels from the liner *Mauretania* to the aircraft carrier HMS *Ark Royal* and the supertanker *Esso Northumbria*, but we discovered that the humble *Lizzie and Annie* also deserved her place in maritime history – not for any great deeds but for simply being one of the unluckiest ships ever.

In her many years trundling up and down the North Sea under both sail and engine power, this iron-clad, ketch-rigged ship, weighing in at just 117 gross tons had been in endless trouble. Even that bible of the maritime world, Lloyds Register of Shipping, described her career as being 'chequered' when they devoted two pages to her in an official publication. And this is why: The *Lizzie and Annie* went aground three times, was in collision seven times, caught fire more than once, broke down too many times to record, was machine gunned by the Luftwaffe during the Second World War and even sank in Whitby harbour, though she was salvaged and put back to work.

She had been repaired so many times that hardly any of her original hull remained but amazingly she was still in one piece, though only just. With Terry's blessing we set off to film the *Lizzie and Annie* in a breaker's yard at Inverkeithing on the Firth of Forth, just days before she was to face her final indignity and be turned into scrap.

We had enlisted the help of a young freelance cameraman, Les Coates, who was both a pal and a colleague, and he agreed to film the story for free unless it was actually shown on TV – quite a gamble on his part as he was working with a couple of novices. I wandered around the ship's rusting, ungainly frame and quaint little aft-mounted wheelhouse telling her unfortunate life story as she lay at a melancholy angle in the mud.

David and Les got some wonderful shots to match the words and my final thoughts were: 'The ship that has refused so many times to die is going to be chopped up into little pieces. But if I were you I'd watch out for the ghost of the *Lizzie and Annie*. Some of those mischievous little pieces might end up in your new car, a razor blade or a can of beans. Who can tell what would happen then?'

And who knows what would have happened to our ambitions if Terry had not liked the film. We anxiously awaited his verdict in the editing room and to our great relief it was a thumbs-up. It proved to be a poignant tale for our *Look North* viewers in an area with such a long connection with ships and the sea.

So we set ourselves an even more ambitious project – 'recreating' the Great Fire of Newcastle of 1854, caused when a sulphur warehouse on the riverbank exploded with a force that shook the whole of Tyneside. What a challenge to capture the horror of that catastrophe with some old sketches and paintings, some imaginative camera work and a little experiment with six pounds of sulphur. In the warehouse that night there had been 2,800 tons of it; huge blocks of masonry hurtled across the river and both banks were in flames.

More than sixty people died as burning rubble rained

down onto their slum homes in what, at the time, were the most densely populated streets in the country, and eight hundred were injured. Queen Victoria stopped the royal train on her way back from Balmoral to survey the damage and express her sympathy.

In the aftermath a proper fire brigade was set up, slums were replaced by better housing and warehouses ordered to take more care over the storage of chemicals but, as we reflected, it had been a hell of a price to pay for progress.

David and I continued to make little feature films and gradually I was allowed to go out as a news reporter. One of my first interviews was with the Bishop of Durham, the Right Reverend Ian Ramsey, and my cameraman on that day was a lovely old character called Arthur Nicholson. To make sure his subjects were in focus, Arthur would mark out a cross on the ground where they had to stand and then use a tape measure to get the precise distance from his lens. Before the bishop arrived I stood in for him on the appointed spot and when he did appear Arthur said to him, without really considering his words: 'You see that cross there, Bishop – do you mind getting on it?' To which the bishop replied with a wry smile: 'Well, there is a precedent.'

We were a very sociable lot at BBC Newcastle and spent many a happy hour after work in the Portland pub next door, where the formidable barmaid once told a visiting boss from down south who had the temerity to order a gin and tonic with ice: 'We only have ice between November and March and that's on the pavements.'

It was a small, rather basic and often very busy pub so we decided it would be a good idea to start our own BBC Club, where we could have more privacy. The management gave us a room in Crestina House, where the newly launched BBC Radio Newcastle was based, a mile or so from our studio. The snag was we did not have enough club members to pay for staff so the committee – me included – had no choice but to volunteer to man the bar on a rota basis.

We soon got the hang of pulling pints and even had ice for anyone who wanted G&Ts. No one expected the service to be perfect and the occasional wrong order or under/over-charging quickly got sorted out and just added to the pleasure of having our own retreat. One October evening in 1968 I was on duty alone on the news desk, but during my supper break I nipped over to the club to work behind the bar.

The phone rang and it was Bert Wilson, our freelance correspondent in Durham, who told me he was getting rumours of a very big story – one that sounded almost unbelievable. He had heard from a contact that some prisoners had escaped from the maximum-security E-Wing of Durham Jail, the supposedly impenetrable prison-within-a-prison where all the hard cases were locked up.

I quickly alerted the national news desk in London, handed over the club keys and dashed back to the office. Less than an hour after pulling pints (but luckily, as it turned out, not drinking them) I was live on the *Nine O'Clock News* from the *Look North* studio. By then I had learned that three men were involved in the breakout, and though two may have been caught while still inside the jail, the other had scaled the walls and was on the run.

It was my first appearance on national news and everything was happening so quickly. One moment I was on the telephone to the police, the next moment the red light was on and the newsreader in London was handing over to me.

1. My parents on their wedding day, 1937.

2. Dad, far right, at the Abbey Stores.

3. Just before Dad went to war, 1941.

4. Me as a baby, just after Dad went away to war.

5. Mum and me before Dad came back from the war.

6. Penny for the Guy! (I'm third from left in the front row).

7. Mum, Jean, Dad and me at Bridlington, c.1949.

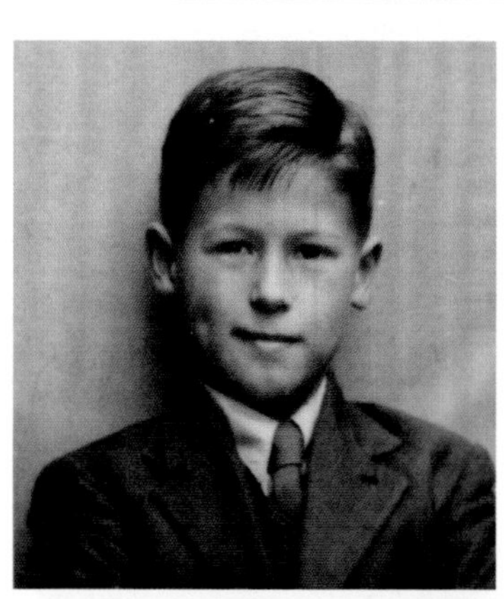

8. The budding journalist. Me, aged about ten.

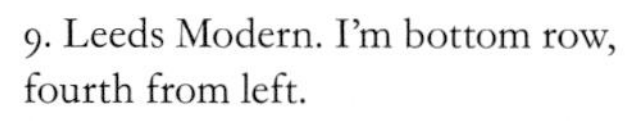

9. Leeds Modern. I'm bottom row, fourth from left.

10. Jean and me. She would have been about three or four years old.

11. Mum, Jean and me.

12. Jean and me on my BSA Bantam 125, 1957.

13. A young freelance reporter in Bradford, 1962.

14. Looking dapper as Niccolo, with Nicolette, Leeds Children's Theatre, 1958.

15. Hard at work at the *Harrogate Herald*, 1959.

16. The famous bubble car. It's more spacious than it looks, honestly, and quite stable!

17. Out and about in Wharfedale c. 1959.

18. A publicity shot for *Newsround.*

19. A photoshoot for a launch of a new season of children's programmes.

20. Learning to play the Hmong pipes at a refugee camp in Thailand in the late 1970s.

The British Academy of Film and Television Arts

CERTIFICATE of NOMINATION

THE COUNCIL HEREBY CERTIFIES THAT

JOHN CRAVEN and JILL ROACH

NEWSROUND EXTRA

WAS NOMINATED FOR A

BRITISH ACADEMY AWARD

FOR OUTSTANDING ACHIEVEMENT DURING 1975
IN THE CATEGORY

REDIFFUSION STAR AWARDS - CHILDREN'S ENTERTAINMENT

CHAIRMAN

ADMINISTRATOR

21. Recognition from BAFTA.

All very nerve-wracking but exciting, and what followed was an extremely long shift, reporting developments during the next twelve hours for television and radio news, the World Service, our own early morning bulletins and ending with the *Today* programme.

It became clear that the three men – John McVicar, Walter 'Angel Face' Probyn and Joey Martin, had escaped from E-Wing's shower room down a ventilator shaft and then got into an exercise yard. Martin was quickly captured and Probyn, who had escaped sixteen times from various prisons – 'it's like a hobby', he once said – gave himself up before reaching the outer wall.

But McVicar, an armed robber who became Britain's most wanted man, made a clean getaway through the streets of Durham – a city he didn't know. For a while he hid in freezing water along the banks of the River Wear, hearing police dogs close by, before evading the huge manhunt and making his way back to his old haunts in London.

A reward was offered and after being free for nearly two years he was recaptured. Back in prison he changed his ways, did A levels and following his release in 1978 became a respected writer. Later a movie was made about the escape and his life on the run starring Roger Daltrey from The Who as McVicar and Adam Faith as Walter Probyn.

Personally, I'm rather grateful to McVicar and co. I couldn't have hoped for a better breaking (or perhaps that should be breakout) story to launch my national news career and as well as getting a letter of congratulations from the head of BBC North there was some extra cash in my pay packet – more than enough to buy a large round of drinks back in the club bar.

During my career I have seen many upsetting moments in court but none more so than in December 1968, when an eleven-year-old girl, Mary Bell, was found guilty at Newcastle Assizes of the manslaughter of two young boys.

She herself was so small that we could hardly see her in the dock – her head peeked above the top as she faced the full panoply of our bewigged legal system. A thirteen-year-old playmate named Norma was charged with her but acquitted. Mary Bell was said to have strangled the boys, aged three and four, 'solely for the pleasure and excitement of killing'. One was found dead in a derelict house, the other on waste ground in the inner-city suburb of Scotswood.

Mr Justice Cusack, sentencing her to be detained at Her Majesty's Pleasure, said there was 'a very grave risk to other children if she is not closely watched'. She was sent to an approved school in Lancashire – the only girl among two hundred boys. She was later detained in an open prison until her release in 1980, when she was given a new identity and she is now a grandmother.

My years on Tyneside helped form me as a journalist and I became an 'adopted Geordie'. I loved the people, their humour, their friendliness and their accent. Mine had been somewhat flat after an upbringing in Yorkshire, but after only a few months my voice started to pick up a little of those distinctive up-and-down inflections.

The social life was great – the north-east is a very welcoming place and largely through work I made many friends. Early on during my time on Tyneside one of them became my girlfriend. We got engaged and married but unfortunately it didn't work out and after a couple of years together we divorced. Both of us went on to have a lasting relationship with a new partner.

David was a great support after the break-up and, together with Tony Bannister, the young graphic artist on *Look North*, and researcher Bernard Hall, who was later to work with David on his cookery shows, I shared a house for a couple of years. Another colleague compared it to an 'upmarket student squat' where we had a lot of fun and beer, and among my more vivid memories is one of Tony sleepwalking through a glass door. Sadly both David and Tony have now passed away.

It was during a boozy night in the bungalow that Neil Armstrong walked on the moon. I was determined to stay awake until the early morning to see the drama of Apollo 11's final approach and landing but the others opted out at around 1 a.m., saying they would watch the replay. How unimaginative, I ranted, when history was being made, but they took no notice. Like millions of others who stayed the course, I was amazed by the fantastic spectacle that NASA put on – extra-terrestrial television at its best, and from my lonely chair in Fawdon I was reaching towards the moon.

The astronauts left behind, of course, that famous plaque: 'Here men from the planet Earth first set foot upon the Moon July 1969 AD. We came in peace for all mankind.' They had signed it and so too had a man who was to become the most discredited leader of the free world. How inappropriate that the name Richard Nixon lives on in outer space.

Back down on earth, my bosses at *Look North* decided to send me on a two-week reporters' course at the headquarters of BBC Television News, Alexandra Palace in London. This vast Victorian building at the top of one of London's highest hills was where television was born in Britain back in 1936, but when I arrived it was being used almost exclusively by BBC news.

Film reports arrived from across the world, telex machines chattered out the latest stories and I accompanied reporters on their assignments. I also got to meet the newsreaders – Michael Aspel, Richard Baker, Kenneth Kendall and Robert Dougall, all of them legends.

One of the training sessions dealt with how to operate in trouble spots; it came with the warning not to over-egg a story. We were shown a classic clip (it was never broadcast) of a famous correspondent in a foreign war-zone whispering to camera that a curfew had been imposed. Anyone seen out of doors was likely to be shot, he said, and fear stalked the streets. All the while, and unseen by him but not by the camera, a little old lady with her donkey walked down the alleyway behind him. She stopped to chat to a couple of friends and some children came out to play, totally contradicting his powerful words.

The embarrassingly clear message was that reporters never know what's going on behind their backs, so be careful what you say. We had instructive sessions on editing and scripting news reports and I decided that, as three words represent one second of air time, I would write my reports using three words to the line, so it was easy to add up how many seconds were being used – and seconds are vital in any broadcast. It's a technique that I later introduced to writers on *Newsround* and it caught on.

One of my tasks was to make a short, self-contained news film on a subject of my choice and I racked my brains to think of something that, as a stranger in the great metropolis, I could tackle. Then I remembered that a musical play set in a mining community in the north-east and called *Close the*

Coalhouse Door had been a big hit when it was first performed at the People's Theatre in Newcastle a few months earlier.

Written by the distinguished northern playwright Alan Plater, with music by Alex Glasgow, one of the presenters of radio's *Voice of the North*, the play had now transferred to the Fortune Theatre in the West End. A member of the original local cast was in the London production and so my idea was to interview him about the show moving south and report on how London audiences were reacting to this gritty northern story (not too well, as it turned out).

I got the go-ahead and filmed late one afternoon, doing a piece to camera outside the Fortune explaining where I was and linking to the interview that took place backstage. Everything went to plan except that when I came to edit my story the following day I couldn't find the processed film cans. They had totally disappeared, ominously like that obituary report on my first day on the *Harrogate Advertiser*.

With nothing to edit and no time to make another report I was none too pleased, especially as passing the course was dependent on producing a good film. That evening I telephoned the *Look North* office to explain my predicament and the first thing the producer said was: 'Thanks for the film, John – it was great . . . a London angle on a local story. We showed it on tonight's programme!'

So that's what happened to my film. It had by mistake been shipped to Newcastle overnight and, because it needed no commentary from me, all *Look North* had to do was cut down the interview and transmit it. When I explained to the course leader that he couldn't assess my work because it had already been broadcast, he said he really had no choice but to pass me.

*

Something strange, almost sinister, happened along the north-east coast late in May 1968. Without warning the incoming tide turned red and shortly afterwards seventy-eight people were taken to hospital. Their fingers, hands and mouths felt numb, they were vomiting and some had a peculiar sensation of floating. What could be the connection between these symptoms and the odd goings-on out at sea? Our newsroom went on full alert and pretty soon we had the answer: it was locally gathered mussels.

Most of the victims had bought prepared mussels from a retail outlet while a few had cooked theirs at home. They were suffering from a rare type of shellfish poisoning known as red tide – so rare that only ten cases had been reported since 1828.

Look North needed an expert who could explain more about this phenomenon and maybe give some reassurance. Our producer, John Exelby, knew just the man. John and his wife Judith Hann (who was then science correspondent of the *Northern Echo* and went on to become one of the stars of *Tomorrow's World*) knew him from their days at Durham University, where he was an enthusiastic lecturer in botany.

The year before, he had commented in public about the impact on sea life of Britain's biggest-ever oil spill after the tanker *Torrey Canyon* was wrecked on the Seven Stones reef between the Isles of Scilly and Cornwall. But would this man be available today, did he have in-depth knowledge of the red tide and was he willing to be interviewed on television?

A phone call confirmed all three and I set off with a film crew to meet him on the Durham coast. By now, the sea was back to its normal grey but my interviewee turned out to be really colourful – an affable, bear-like man with what looked like a broken nose and a passion for his subject which he expressed in the most distinctive voice.

He explained that the red tide was caused by a massive blooming of algae which discoloured the sea and made mussels toxic when they ingested it. In short, he gave a brilliant, fluid and reassuring interview and I am sure you will have guessed by now that his name was Dr David Bellamy.

The danger passed, all the victims recovered (in other such incidents there had been deaths) and by the end of that summer mussels were back on the menu. But I think it's fair to say it was that interview that launched David's television career.

He went on to make more than 400 programmes and turned botany into prime-time viewing. As an environmental campaigner he went to prison in Australia for protesting against a proposed dam, and claims he is now shunned by TV because he believes global warming is 'poppycock'. The last time we met was at a lecture in Buckingham Palace to mark his eightieth birthday and together we recalled that meeting we had along the Geordie shores which opened up a new career for him.

For me, BBC Newcastle was a perfect training ground where I could practise my skills under the guidance of people I admired and trusted. And I even got to invent a quiz show when the search was on for a tranche of diverse programmes to fill a late-night, thirty-minute slot on BBC One.

Our region stretched from the Solway Firth in the west to Wallsend in the east, so what device could we use to link our viewers in a fun way?

The answer to me was Hadrian's Wall, and we built a tiny replica of it in the studio for a quiz we called The Wall Game. Instead of getting points for every correct answer the two teams moved to the next milecastle along our wall, or rather their toy-sized chariots did. Mike Neville was centurion-like as the quizmaster and at least it looked different from all the other tests of general knowledge. Just one thing distinguishes the real wall from my quizzical one – it is still there!

Nightclubbing was big in the 'party city' of Newcastle. At one time in the 1960s there were no fewer than eleven clubs and the biggest, most glamorous of them was La Dolce Vita. It was named after the Federico Fellini film starring Anita Ekberg that focussed on the 'sweet life' in Rome, full of pleasure and indulgence.

Political leaders in the area had talked of Newcastle becoming 'the Venice of the North', so there was an Italian connection, but the political dream was totally different from that of three local brothers, the Leveys from Wallsend, who owned La Dolce Vita. They were bringing what they hoped was movie-like glamour and Las Vegas-style entertainment to a north-east which was still shaking off its cloth-cap image but had stepped wholeheartedly into the Swinging Sixties.

Suddenly Newcastle, still earning much of its living from shipbuilding and heavy engineering, was the place to be if you wanted to be a little glitzy, sip a cocktail and watch top-class international entertainment. It didn't cost a fortune at the door as money from the gambling tables helped pay the bills.

The club had seating for four or five hundred, a glass dancefloor that was lit from below and a stage which hosted some of the biggest names in show business . . . American stars like Louis Armstrong, Sarah Vaughan, Mel Tormé, Billy Eckstine and Jerry Lee Lewis would fly over to appear in London then head to Newcastle for a week.

I interviewed, very briefly, waning Hollywood sex symbol Jayne Mansfield, who didn't seem to know quite where she was during her tour of nightclubs in northern England and who, three months later, was killed in a car crash in Mississippi. British headliners at La Dolce Vita included Tom Jones, Shirley Bassey, Cilla Black, Helen Shapiro, Manfred Mann, Tommy Cooper – the list went on and on.

Bob Monkhouse was brilliant – a comedian at the top of his game with a side-splitting series of one-line gags. But the young David Frost bombed when coming face to face with a wisecracking Geordie audience that took no prisoners. Years later Roy Castle confessed to me that when he appeared at La Dolce Vita he made the mistake of saying he'd just appeared at the Palladium.

From somewhere in the crowd came the shout: 'Where's the Palladium?' Then, as Roy stood there helpless, the audience took over. 'It's in London,' cried a voice in the darkness. From elsewhere: 'Where's London?' From another corner of the room came: 'It's south of Sunderland.'

Almost inevitably there followed: 'Where's Sunderland?' And, after a perfectly timed pause: 'Second from bottom.' It was a classic example of Geordie humour coming from the seats, not the stage, and taking a swipe not just at the north–south divide but also at the eternal rivalry between the Newcastle and Sunderland football teams. And the audience loved Roy when he took charge again and showed off his great talent as a musician and comedian.

Not all the famous visitors to La Dolce Vita went on stage. There were reports that the Beatles called in for a drink after a concert at the City Hall, as did Ella Fitzgerald. But how sinister was it that former world boxing champion Joe Louis was photographed there with a couple of infamous guests, the Kray twins Reggie and Ronnie. The appearance of these

London gangsters sparked fears that the underworld was trying to infiltrate the crime scene on Tyneside.

Those fears were heightened when, late one night in January 1967, a man named Angus Sibbet, whose job was to collect money from 'one-armed-bandit' gambling machines, left La Dolce Vita and next morning his body was found in his red Jaguar under a bridge some miles away. He had died from three gunshot wounds in what appeared to be a gangland execution. Two men, one a Cockney gangster, were jailed for his murder and the circumstances surrounding the Sibbet case are said to have inspired one of Britain's best-ever movies, *Get Carter*, starring Michael Caine as a London hard-man out for vengeance on Tyneside.

It was a violent, uncompromising film made on location in and around Newcastle in 1970, and though *Look North* tried many times to get an interview with Michael Caine he was always too busy. Then one morning the news desk got a call from the production company saying they were shooting a sequence in the Long Bar, a tough drinking den in the centre of town, and if we could be there at about lunchtime we could have an interview.

As all the reporters were out filming, it was left to me to take on the best job of the day. We did some shots of the movie crew filming Michael Caine in the dimly lit bar and then I had my chat with the star. This was the first time I had done a 'celebrity interview' on television and he could not have been more helpful. We leant against the bar counter and he gave me the background on the tough role he was playing and his impressions of Newcastle. When we had finished we decided there was time for a quick drink.

'What'll you have, Mr Caine,' asked the barman. 'Well, as

it is a little early in the day I'll just have a half of lager,' he said. To which the barman replied: 'Sorry but we don't have any lager.' 'Why ever not?' asked the bemused star. 'Because,' came the devastating reply, 'we don't get many women in here!'

So here was a top box-office name, playing the hardest of criminals, being told that when it came to alcohol his choice was a little feminine. Don't forget, though, that back then lager and lime was THE ladies' drink, especially Up North, and it would be many years before real men in places like the Long Bar drank lager.

A few days later the film company invited me to a reception at the Royal Station Hotel, where lager was on offer. Michael Caine made an eye-catching appearance with his beautiful Guyanese girlfriend Shakira, later his wife. Tinseltown had come to Tyneside.

And La Dolce Vita? The end began when it opened a disco on the floor above (entrance price: girls five shillings, guys a guinea) and after changing its name several times the site was turned into flats. The sweet life was over.

But not for me, because I had fallen in love again – and this time I knew in my heart that it would be for keeps. Marilyn was from Cleadon, a village between South Shields and Sunderland, and, having returned from a year working in Australia, she was now a production secretary on *Look North*. She was lovely, vibrant, strong in the way that north-east women are, and all I had been looking for.

Shortly after we became engaged, our love was put to the test. Pat Morley, who had briefly been my boss in Newcastle, had been made news editor in Bristol and was looking for a full-time freelance reporter for *Look North*'s sister programme *Points West.* After six fantastic formative years, should I apply?

Would Marilyn, recently returned to the north-east after a year away from home, be prepared to move again? The answer to both those questions was 'yes', so we agreed that if I got the job she would look for work there as well and we would marry in Bristol – a new chapter for both of us.

6. Bristol

I caught an early morning Dan-Air flight from Newcastle to Bristol for the big meeting that would decide my future. It was a beautiful, cloudless July day in 1970 with fantastic views down below, but suddenly that future seemed a little less promising when I spotted that a needle on the fuel tank beneath the overhead wing was pointing to DANGER.

I watched it closely for a while, hoping it would switch back to normal, but it didn't. With no other passengers close by to share my concern, should I warn the captain or was I worrying about nothing? Finally, I pressed the call button. When I showed the flight attendant the needle she smiled sweetly and said: 'Don't worry. That always seems to be happening – I'll let the captain know. But sit back and relax, we'll get you there safely.' She was right, but could it be one reason why the airline got the nickname 'Dan Dare'?

The job interview with Pat Morley and the West Country regional manager Jimmy Dewar seemed to go well at the BBC studios on the interestingly named Whiteladies Road (at the foot of Blackboys Hill so maybe, I supposed, a reference to the city's slave-trading past). As I was being shown round the newsroom, the local researcher for *Nationwide* took me to one side. 'I hear you've flown down from Newcastle today,' he said, 'and by coincidence we're doing a piece on the programme tonight about internal air travel in the UK. Could you contribute to that?'

I had to think quickly because no one in Newcastle apart from Marilyn and John Exelby knew I was there and I didn't want my colleagues Up North to find out about my job interview via prime-time TV – especially if I didn't get it. So I said that unfortunately I was too busy to record a report and could not do it live because at that time I would be back in the air again. Narrow escape!

After a quick tour of the city that could be my new base I flew back to Newcastle on the same plane, not even bothering to check whether that needle was still on danger. The next day I got a letter from Jimmy saying I would be a welcome addition to his team. 'I realize this will mean resignation from the BBC but you must decide between the security of a staff job and the chance to do work that you prefer,' he said. 'I am preparing a contract to offer you and when you have thought it over please let me know what you want to do. We hope you will join us as quickly as possible.'

It was a no-brainer, with Marilyn to be shortly by my side, though I did get a hint there might be a move up the staff ladder for me on one of the new local radio stations. But television was the medium I wanted to be part of so, saying a sad farewell to all my good friends in Newcastle, I took a chance that I was not to regret.

Reporting from the West Country turned out to be a totally different experience from the north-east. People were more restrained, the countryside was gentler, the accent less sing-song and the economy more affluent – but the welcome was just as warm.

During what was to be four years on *Points West* I filmed nearly 600 reports and my 'patch' was a big one covering Gloucestershire, Somerset, north Devon and parts of Wiltshire. Every day I was sent out on a story for that night's programme and had to be back with my film in the editing room by around 3 p.m. The cameraman was issued with 400 feet of black-and-white film that had a sound track running down the side and a hundred feet of mute film that would be laid over the main film to illustrate the story.

The reel of sound film lasted for ten minutes and the finished story was expected to run for no more than three minutes and eight seconds, which is a ratio of roughly three to one. And we couldn't ask for more film because it was expensive. That called for a lot of discipline when you were gathering your report on location. Questions needed to be carefully prepared, interviews kept short and interviewees persuaded to keep to the point.

Everything had to be concise yet fluid. Reporters in today's digital age are not as restricted in terms of the amount they can shoot and they don't have the nightmare of being stuck in a traffic jam on the A38 miles from the studio with the minutes ticking away to the deadline and the film cans still in the boot of the car. They can edit their reports on a laptop then send them to the newsroom on the internet.

My films were handed to an editor who would cut the film down to size, with me looking over his shoulder and often pleading for an extra few seconds above the three minutes eight seconds limit, because then I would get paid a bit more. I would dash back to the office through the labyrinth of fine old buildings that made up BBC Bristol and write and record my commentary (or, if we were running late, read it live in the studio).

When *Points West* was on the air my report was transmitted from one telecine machine while, on a time cue from the production secretary in the gallery, another machine showed the mute pictures in a technique known as second telecine overlay.

Every day was different because you'd never know what story you'd be handling, from murders to a horse that liked a pint in a pub. Sometimes I'd be sent on 'safari' for a week in north Devon and told to come back with ten stories with only me to find, research and film them.

It meant browsing through local newspapers and making endless telephone calls to contacts. That's how I came across both the boozy horse (fortunately, the bar he favoured had a stone floor) and an elderly retired actress in a small shabby flat in Ilfracombe who had once appeared in Hollywood films and had, she claimed, dated the movie star Ray Milland.

In those days, long before mobile phones, the only way to keep in touch with the office was by using phone boxes, and they could be unreliable – either out-of-order or occupied. So my white Austin 1300GT with a black vinyl roof (remember those?) was fitted with a version of citizens-band radio so that the office could always be in contact and vice versa.

That, at least, was the theory. My call sign was Silver Charlie Three, which was exciting, but the system didn't always work and when it did I had to spend much of my driving time listening to messages about broken-down one-armed bandits. To save money, BBC West shared the waveband with a firm that supplied and repaired fruit machines in pubs and clubs and their calls often blocked ours. With this compromised form of communication we had hardly hit the jackpot, unlike our airwave partners, but it was better than phone boxes.

After I had been in Bristol a few months I went into Studio

B one Friday afternoon to record my commentary and discovered it had already been booked by people I didn't know. 'What's going on?' I asked one of them and was told: 'Sorry, but you'll have to come back later because we're doing auditions for a current affairs programme.' 'Well, can I have a go?' I asked, and that simple question would – though I didn't know it at the time – herald my return to network television for the first time since *The Sunday Break* in my late teens.

'Actually, we're looking for a presenter for a rather different kind of current affairs show, because it's aimed at children,' said a man who turned out to be one of the directors. 'It'll be the first of its kind and it's going to be called *Search*. We've been auditioning for two days, seen a lot of people and we are almost at the end of both the talent and the videotape. But as we have taken over your studio today, I'll do a deal. If we do have any time left I'll give you a call.'

That sounded like a pretty good excuse for sending me on my way, but I gave him my office number and thought no more of it. After all, the idea of working on a children's programme had never crossed my mind. But some time later the phone on my desk rang and a voice said: 'It's the *Search* programme, John. If you'd still like to audition can you make your way quickly to Studio B as we have about six minutes left on the videotape?' I took off my jacket, loosened my tie to look a bit more casual and hurried down.

'The programme will involve groups of children talking to the presenter in a studio, giving their opinions on big issues,' explained the director, 'so we'd like you to read an introduction on autocue and then chat to these children here.'

He pointed to a small group who had, before the auditions started that day, been primed to be awkward, and after a few hours of being just that they were now positively grumpy. I had six minutes to show what I could do and there was no chance of extra time because just one reel of two-inch-wide videotape had been allocated.

The six minutes flew by. The children perked up and, though they did give me a bit of a hard time, I can't remember what we discussed. The whole, unexpected experience was somewhat surreal and at the end of it I went back to reality and recorded my script for that night's *Points West*.

The audition tape was sent to London to be viewed by the head of children's television, Monica Sims, and her deputy, Edward Barnes; they would have the final say. I more or less forgot about it – after all, there were some talented people on that list and I was the unscheduled latecomer that no one had ever heard of.

'I believe Monica and Edward were getting frustrated going through the numerous takes and thinking the whole exercise had been a waste of time,' Brian Hawkins, a friend who joined the *Search* production team, recalled. 'Finally your piece came up and without hesitation they decided you should fill the position.'

But first they wanted to meet me. I couldn't believe my luck and, looking back on why I did well, it must surely have been the fact that I didn't have any nerves. I hadn't spent weeks worrying about the audition – it was just something that happened out of the blue that particular afternoon.

And a few days later I was standing outside the front gates of BBC Television Centre at White City, waiting for a commissionaire to admit me to the most famous television studios in the world. It was my first visit to the 'dream factory' and I was in total awe. Built in a circle with eight large studios on the ground floor, editing suites in the basement and seven floors of offices, it's one of those rare and distinctive modern buildings that can be immediately identified.

The favourite nickname for it was, because of its shape, the Magic Doughnut and I've also heard it said that Television Centre was built in a circle for one very good reason – so the buck couldn't stop! No doubt there's some truth in that, but on that cold, blustery day in January 1971 I had the strange feeling it was welcoming me. While waiting in the airy main reception my jaw just about hit the ground, because almost every other person strolling past was famous. Maybe that's a slight exaggeration, but this was *the* meeting and greeting place for the great and the good.

I was led to the children's department in the East Tower and introduced to Monica and Edward (when I first joined the BBC it took me a long time to get used to the fact that everyone from the Director General to the tea boy is known by their first name).

They quickly put me at ease and first impressions were that he was friendly, bearded and enthusiastic and she was astute and kindly with a somewhat regal air. The BBC was renowned for its tough grilling of applicants at appointment boards but potentially this was worse. Not only was I being questioned in a reasonably formal way, but there were other considerations to take into account because the questioning continued over lunch in the waitress-service restaurant in Television Centre.

Was I holding my knife and fork properly? Was I behaving in an appropriate manner for the kind of role that was being offered? Would it matter that during the meal I once replied to Monica with a mouthful of food?

Well, I can't have disgraced myself too much because I passed the test. *Search* was mine, and in the following years Monica (who sadly died recently), and Edward were not only my employers and advisers but also my friends.

The first of thirteen episodes of *Search* was due to be recorded in Bristol on a Tuesday early in April 1971, and that could have been a major snag.

For some time there had been a very important event marked in my diary for the Saturday before – Marilyn and I were getting married in the Methodist church just down the road from the BBC. We had bought a three-bedroomed house on a new estate in the little town of Thornbury in south Gloucestershire and had been planning a honeymoon abroad.

When I broke the news she was understanding and supportive, as she always has been ever since, and I have never taken on any offer of work without consulting her and heeding her advice. She has been my manager as well as my wife and best friend for more than forty years. In fact, Marilyn says she's been married to me for more than eighty years because she has always had to say everything twice, whereas I've been with her around twenty years because I've only half listened.

We made a pact that after our wedding we'd have a two-day honeymoon in Cornwall so I could be back in time for the recording and then have a much longer one later that year in Tunisia, which we did.

Search covered topics from corporal punishment to bullying to grandparents, showing first a short film report and then handing over to around thirty children sitting in a semicircle for their views on the subject, with me prompting.

Search also welcomed to the studio experts on the subjects we investigated such as science journalist James Burke, disc jockey Annie Nightingale, mountaineer Chris Bonnington and the Editor of BBC Television News Derrick Amoore.

The show had one of the best signature tunes ever – the opening bars of Led Zeppelin's 'Whole Lotta Love' (hardly appropriate, you may think, for a child audience, but we dipped out of it before the lyrics started), which was also 'pinched' by *Top of the Pops* for its theme.

We recorded an hour or so of discussion, and as the children had been well chosen by our researchers it usually turned into a free-for-all with sometimes controversial pearls of wisdom spilling from young mouths. It had to be cut down to around twenty minutes – not easy in those early days of videotape when the editing was done not electronically but with, of all things, a razor blade, with the chosen sections then glued together. The producers had only one chance to get it right.

Search was recorded in colour in the big Studio A in Bristol, which was also the home of other children's shows like *Animal Magic* with Johnny Morris and *Vision On* with Tony Hart, while *Points West* went out live in black and white from little Studio B. I happily settled into the new split role of children's presenter (despite one of the kids saying on air that they could do the job better than me) while still spending most of my time reporting for *Points West*.

I had the best of both worlds. It was jumpers on *Search* and suits on *Points West*, though I wished I'd been dressed rather more casually when I was reporting for the regional programme from the middle of a soggy field at Worthy Farm in Somerset on Midsummer's Day 1971. In that field the first

Glastonbury Fayre was underway – an event which in the ensuing years was to become world famous simply as Glastonbury.

I was surrounded by happy hippies while wearing a business suit and talking in a newsman's way about sex, drugs, rock 'n' roll and mud. It's a clip that's been shown in just about every documentary about Glastonbury ever since. In hindsight, it does make me seem something of a spoilsport, but the West Country wasn't used to that kind of thing and to prove it my report showed policeman with binoculars 'keeping an eye' on the festival from another field.

About 12,000 people turned up, the tickets were free (how different from music festivals today) and a pyramid-shaped stage had been built. We were told it was one tenth the size of the Great Pyramid of Giza and had been carefully placed along an ancient ley line that linked Glastonbury Tor, Stonehenge and the Egyptian pyramids. The theory was that when the sun shone on the morning of the summer solstice its rays would strike the apex and light would diffuse down the sides.

A load of old tosh, we news cynics thought, especially as it had been raining. But there was mysticism in the air – among other things – and maybe I'm hallucinating now as I write this, but I seem to recall that as dawn broke, the skies cleared and beams of sunlight fell on the pyramid as the American singer Melanie greeted the day and the crowd from the stage. Glastonbury Fayre got off to a magical start and none of us realized we were witnessing the birth of a musical phenomenon that seems to grow stronger every year. As the festival went on other stars like Joan Baez and Fairport Convention graced the pyramid stage and David Bowie was the headliner.

Towards the end, I returned to sum it up for *Points West*:

Though to the hippies the Glastonbury Fayre has been a time of peace and happiness for everyone, there have been certain aspects of it that have disturbed what they call 'the straight society' – that's just about everyone but themselves.

Aspects such as the free love-making, the fertility rites, the naked dancing and most of all the drug-taking. There has been pot and acid – that's cannabis and LSD – but there have been only two arrests and the police say they were quite surprised and pleased about the way the festival had gone.

Yes, that does sound rather headmasterish, but in truth Glastonbury had won me over. It had seemed all rather haphazard, harmless and, in the nicest way, innocent – a gathering of young people living a largely alternative life-style; a mass of kaftans, flares, long straggly hair, muddy feet and bountiful goodwill. Over the years Glastonbury developed into the huge commercial event it is today and I've always intended to go back but never have, perhaps knowing that it would never again be quite like it was the first time.

Just before Christmas that year I was to encounter anything but goodwill when *Points West* sent me to Belfast to cover our local regiment, the Glorious Glosters, which had recently been posted there and was patrolling the streets as the Troubles mounted. Throughout its long history the Gloucestershire Regiment had distinguished itself in battle, from Egypt in the time of Napoleon to the Korean War, but this was different.

This enemy had vowed to get British soldiers off the streets and out of the country with guns and bombs. On my way by taxi into the city centre from the international airport I got a taste of what was to come. Instead of pointing out favourite beauty spots and places of interest like other taxi drivers the world over, mine was keen to show me the locations of the latest terrorist attacks. Welcome to Belfast!

Because the director wanted first-hand footage of soldiers both on and off duty, we took up the offer of staying with the troops at the Hastings Street police station, where they were billeted. We discovered we'd be sleeping on makeshift beds in what had previously been the cycle shed just across the quadrangle from the main building.

'It can be dangerous,' we were warned by the officer looking after us, 'so don't just stroll across – you'll need to run quickly in zig-zag fashion, bent over double.' So where could any danger come from as this was a heavily protected building? 'From over there,' said the officer, looking towards a tower block of flats that dominated the area. 'That's a favourite spot for IRA snipers.' So for three days and nights we zig-zagged across that wide open space, hearts in our mouths.

'You must be crazy,' said the seasoned BBC reporter Bernard Falk when I saw him at an army briefing. 'I'm staying at a very safe hotel and they still have rooms available so why not check in there?' I thanked him for his advice but said we wanted to be close to the Glosters.

A few hours later it was Bernard regretting his choice of accommodation – his hotel had been bombed while he was out, his room destroyed, and he had lost everything apart from the clothes he was standing in. Five firebombs went off within five minutes of each other in Belfast that day, injuring a dozen people as the explosions ripped through the city and sent Christmas shoppers running for cover. As we drove through the deserted main square in the early evening a large banner wishing everyone a Merry Christmas seemed somewhat inappropriate.

Spending time in that frightening atmosphere brought home to me just what soldiers and civilians alike were facing every day. One morning I was walking along the street next to our police station on my way to buy the daily papers when suddenly there was a loud bang. Instinctively I threw myself onto the pavement, only to see feet striding past me. No one else had taken evasive action and it turned out the noise had come from a vehicle backfiring as it crossed a speed hump. I felt rather foolish but consoled myself with the thought that it was better not to take chances when you were a first-timer in Belfast in 1971.

In one sequence for our film we followed the Glosters on foot patrol along the terraced streets of the Lower Falls, the Republican area in west Belfast. The scene confronting me was hard to take in as young, heavily armed soldiers, constantly on the alert, moved past mothers pushing prams and children playing games. It was a weird visual contradiction; ever-present danger and the perfectly normal.

To be honest, when our filming was completed I was relieved to be leaving the city, though it must be said that the terrorist organizations did not attack journalists, believing, I suppose, in the old adage 'Don't shoot the messenger'. But since then I have been back to Northern Ireland many times as a reporter, a holidaymaker and even as a compere introducing concerts by the Ulster Orchestra all around the province, and I have grown to love the place.

During the Troubles I never failed to be impressed by the way people on both sides of the divide refused to be cowed and were determined to go about their normal lives no matter what was happening all around. Hours before a concert in Belfast there was a terrorist attack and I asked the orchestra manager if there would be empty seats that night. Would some concertgoers prefer to remain in the safety of their homes? 'Not at all,' he said. 'When something like this happens people show they will not be intimidated. The hall will be packed.' He was right.

But that first visit in 1971 is still vivid in my memory, especially as, when we were leaving the city, the taxi driver completely ignored the sign to the airport and set off towards the IRA stronghold of Londonderry/Derry. Were we being kidnapped? 'Didn't I just see the airport sign back there?' asked our director, keeping his cool. 'Oh, sorry about that,' said the driver, 'I was daydreaming.' We breathed sighs of relief and caught the plane home.

I continued to work for both 'grown-up' news and children's programmes throughout my years in Bristol. *Search*, always an innovative show, launched an annual film competition in which children used their own or borrowed 8mm movie cameras (no home video cameras then) to produce documentaries, mini-dramas or cartoons.

These short films required imagination as well as technical ability and the standard of many of the entries was incredibly high. As their prize, the winners got to make a full-blown short film for the programme using BBC film crews and it was quite a sight as highly qualified cameramen, sound recordists and film editors (some of whom worked for the BBC's Natural History Unit and had won worldwide plaudits) took their instructions from a keen bunch of youngsters.

Many years later I was reminded about the *Search* competition by two well-known figures – the television presenter Trisha Goddard and the creator of Wallace and Gromit, Nick Park. We three met at a reception for the media given by the Queen and other members of the royal family and as we waited for them to reach our little group Trisha said to me: 'I doubt you'll remember this, John, but when I was in my early teens I made a film for the *Search* competition and I was one of the winners. I had a wonderful time working with the professional crew and I decided there and then that a career in television was what I wanted.'

Then Nick, the winner of four Oscars for his amazing animated films, chipped in. 'That's quite a coincidence,' said the man who was to find international fame with movies like *A Close Shave* and *The Wrong Trousers*, 'because the *Search* film competition inspired me as well. I borrowed my mum's Super 8mm camera and sent in a stop-start animation film, one of my first. But I didn't win!' Obviously, though, the disappointment didn't put him off. Thank goodness.

Nor did it put off Simon West, who went on to direct Hollywood movies like *Lara Croft Tomb Raider*, *Expendables 2* and *The Mechanic*. When I met him by chance recently he told me: 'I was fifteen when I sent in my film to *Search*, and it was just lots of cool shots of a motorbike riding across country. I can see why it didn't win, but I suppose it was a foretaste of the kind of action films I would make later. And it taught me a big lesson – it's important to have a good story as well as lots of flashy stuff. I still have that *Search* film on a shelf at home.' I've often wondered how many other children who would later make their names in film and television tried out their skills by posting their first films to *Search*.

On another tack, I presented and helped devise with Brian Hawkins a quiz show for children called *Brainchild* which involved Beryl, the acronym for Brainchild Electronic Random Year and Letter indicator, a fake computer. Beryl fed questions to the young panellists and one of the rehearsal rounds was all about the countryside. In answer to 'What would you expect to find on an arable farm?' an eager young panellist pressed the buzzer and said 'Arabs'.

My first big documentary series for BBC children's television was produced by a wonderful lady called Molly Cox, who had been one of the pioneers of the storytelling favourite *Jackanory*, so obviously wrote lovely scripts. Molly was a firm believer that television should broaden children's minds and instil in them a thirst for knowledge. She called our series *The Story Behind the Story* and it examined the truth behind some great historical legends such as the lost city of Atlantis and tumbling walls of Jericho.

It was the first time I had been employed in the role of storyteller and found myself with huge wedges of Molly's words to memorize as I wandered round the settings of these legends. And, for the one and only time during my entire BBC career, I didn't have to provide my own clothes – Molly took me to the local branch of Austin Reed and kitted me out for the series.

And so it came to pass that I was scrambling amongst the dusty excavations of old Jericho, close to the banks of the River Jordan, retelling the biblical tale of how the walls of the Canaanite city collapsed after the Israelite army marched round them for seven days carrying the Ark of the Covenant. There seemed to be little archaeological proof to back up the Old Testament story, and in fact some experts thought Jericho was a pile of ruins long before Joshua wrote about it.

While going through the next sequence of the script with Molly I spotted our cameraman, John Else, searching in the rubble. 'I've just found my tripod marks from the last time I was here, filming *The Ascent of Man* with Jacob Bronowski,' he said, casually mentioning one of the greatest TV series of all time, when the Polish-born professor traced how mankind developed through our understanding of science.

'There was a shot I failed to get because the light wasn't right. But it's perfect now, so if you could position yourself just there, where the professor stood the last time, I can get it.' And he did, with me literally in the footprints of a genius.

Later we flew from Israel to Turkey to examine the myth of King Midas and at Lod Airport in Tel Aviv every item of luggage, from camera equipment to clothing, was thoroughly checked by security guards. Tension was still high after the Arab–Israeli War of 1967 and the October War of 1973 was, as it turned out, only a few months away.

Halfway through the flight, on which we seemed to be the only Europeans, John looked out of the window, then turned and said: 'Something is wrong. We should be flying directly north and the sun is in the wrong place. We must be heading back to Israel.' No sooner had he said that than the captain was on the speakers with a few words in Turkish and fellow passengers began shouting and screaming as the airliner suddenly plunged towards the sea.

We learned later than an anonymous caller had warned the airport in Israel that an altitude bomb had been put on board and the captain was dropping to what he hoped would be a safe height. After a few minutes we spotted Israeli jet fighters just beyond our wingtips escorting us to Lod. Some form of calm returned to the cabin as the aircraft levelled out and once we were safely on the ground every item of luggage was placed on the tarmac.

Everyone had to identify their own cases and as we were driven to the terminal I noticed that one solitary suitcase remained behind and was being loaded onto a military truck. Despite our questioning, no information was released. We never discovered whether there had been a bomb on board, and several hours after we should have landed in Turkey we were ushered back onto the same plane and resumed our journey.

But for me the drama wasn't over. As we went through the X-ray check in Turkey I was brutally and without warning pushed away from my colleagues and into a side room by a squad of customs men and armed soldiers. Something in my rucksack was creating the panic. Then I remembered that in a souk in Jerusalem I had bought a couple of brass eggs as a souvenir. They must have shown up on the X-rays like hand grenades.

One of the soldiers spoke English and I explained to him what had happened and showed him my BBC identity card, to prove I wasn't a terrorist. He allowed me to dig the eggs out of my rucksack and everyone relaxed. I was sent on my way with a stern warning. Some journey!

Another *Story Behind the Story* took me to the fascinating Greek island of Santorini in the Aegean Sea. We missed the ferry from Crete, seventy miles to the south, but were told a cargo ship would soon be leaving for the island and our attractive young researcher offered to see if she could hitch a lift.

The ship's officer on duty on the quay was only too happy

to oblige. 'Can I bring some friends along as well?' she asked. Hoping they would be as pretty as she was, he nodded with a smile, which soon disappeared when he saw the rest of us emerging from behind a shed – but he kept his word.

As we approached Santorini we could see the white-painted houses of its main town perched on top of sheer, 300-metre cliffs. The pathway from the sea to the town was so steep the only way to get our equipment up was to hitch another lift, this time on the backs of donkeys.

Santorini's donkeys spend their lives trudging up and down that pathway carrying supplies and tourists from cruise ships. It's such hard work for these beasts of burden that donkey owners all over Greece apparently threaten them with the words: 'If you don't work harder I'll send you to Santorini.'

From the top it's even more obvious that the island and its cliffs are all that is left of a huge volcano which exploded 3,600 years ago in one of the biggest eruptions the planet has ever seen. It sent a tsunami crashing into the island of Crete, destroying the Minoan civilization there. Could that have been where the Atlantis legend came from?

We spoke to the team of Greek archaeologists who'd been unearthing evidence left by people who lived there until the eruption and some seemed convinced that it was. On our last night they invited us to a barbecue.

During the return journey in a minibus they started to sing songs by Mikis Theodorakis, the composer and song-writer best-known worldwide for his music for the film *Zorba the Greek*. They were halfway through a haunting love song when suddenly they stopped. 'Please carry on – it's beautiful,' I urged. 'We can't,' came the reply, 'because we have reached the town and the police might hear us. The music of Theodorakis is banned throughout Greece and it is an offence to play or listen to it.'

The words sent a shudder down my spine because how could it be, in this cradle of democracy, that love songs were banned? But it was in the time of the Colonels, when a right-wing military junta ruled Greece and Theodorakis was seen as a political threat and banished along with his songs. He came back from exile after the fall of the Colonels in 1974 and thankfully his music returned with him.

Those were just some of the stories behind *The Story Behind the Story* and Molly and I were to work together again some years later on another documentary series, *Breakthrough*, about the lives of great scientists and engineers – the likes of Isambard Kingdom Brunel, Thomas Telford and Edward Jenner, the Gloucestershire doctor who pioneered vaccination. They were serious but very visual thirty-minute programmes of a kind you don't see on children's television today.

It was early in 1972 that I got an offer that was to change my professional life – and it was thanks to Jonathan Dimbleby.

7. *Newsround*'s First Years

'Hello again.' That was how I greeted the nation's children more than 3,000 times as I welcomed them to *Newsround*, their very own television news bulletin, the first in the world. But what I didn't know until recently was that I wasn't the first choice for the job. 'I had you at the back of my mind,' said Edward Barnes, the BBC executive who created *Newsround*, when we were reminiscing about the early days, 'but I was treading carefully because you were bound up with *Search* and I didn't want to be accused of stealing you. So I asked Jonathan Dimbleby if he would like to present it. But a day or so afterwards he got a job on the ITV version of *Panorama* and mistakenly he thought it was a better offer!' How about that – instead of being John Craven's *Newsround* it could have been Jonathan Dimbleby's. Mind you, his career has turned out rather well and it was certainly a lucky break for me. Jonathan said recently: 'It is true I was offered the job and turned it down. The only programme that I wish I had been offered but never was is *Countryfile* . . . I fear it is too late now.' If only I had known that years ago, Jonathan, maybe I could have returned the favour . . . or maybe not!

The man who sealed my fate was not Edward but the editor of BBC Television News, the charismatic if unpredictable Derrick Amoore who was one of the greats in the early days of current affairs programming, had founded *Nationwide* and was once described as being 'brilliant but bonkers'.

If someone today had the idea of setting up a news programme for children it would never see the light of day for the simple reason that the costs would be too great – many millions of pounds just to get it off the ground. But back in 1972 the mood was different – funding could also be tight but there was a freewheeling 'let's give it a go, we'll find the money somehow' attitude which gave birth to many pioneering shows, including *Newsround.*

So how did it begin? Edward recalls Monica Sims, who was in charge of children's television, issuing a press release saying that her department was a microcosm of adult programming. It covered the entire spectrum from drama to wildlife to light entertainment – in fact, everything apart from news. The reason for that was simple – most children simply hated the news. It brought an abrupt end to children's programmes at 5.40 in the afternoon and it involved a man in a suit telling them things they didn't understand. It was all talking heads and boring and only a tiny fraction of children watched it. Edward set about changing that perception.

But it wasn't the first time the BBC had attempted to give the younger audience a taste for the news. During the 1950s there was *Children's Newsreel*, a fifteen-minute programme once a week which was an offshoot of the adult *Newsreel.* The one difference was it didn't have any real, hard-hitting news – it was full of maypole dancing and other ancient customs, craftsmen making cricket bats and even a stationmaster's dog taking a day trip to the seaside on a steam train, all accompanied by a jolly tune called 'Holiday Spirit'.

You got the impression that if a news cameraman saw something quaint happening while he was hanging around waiting for a big story, he would shoot it for *Children's*

Newsreel. One of its films, though, from 1952, became part of television history. The producer Don Smith made a very different version of a steam train trip to the coast – no dog this time, but with the camera speeded up and the journey taking just four minutes, the equivalent of travelling from London to Brighton at 765 mph.

The film was later shown on the adult *Newsreel* and has often been repeated over the years. Edward wanted *Newsround* to be a real bulletin – not children's news but news for children – which is a fine but vital difference – and the idea grew from the very prosaic need to fill a gap in the afternoon schedules.

Part of his job was to ensure the children's programmes segment ran to time and he found that very often there was a five- or ten-minute hole in the schedules. 'If you put in a not-very-good filler that was a great way of handing over the audience to ITV,' says Edward. 'Then I thought of an elastic-sided news bulletin that could fit into any time that was spare. That's really how it started. I talked to Paul Fox, who was controller of television, about the idea and he was very supportive and said I should go straight to Derrick Amoore.'

What followed was one of those surreal but effective encounters that could only happen in the days before accountants with concepts like total costing took over the purse-strings of broadcasting. Edward was leaving the canteen one lunchtime when he spotted Amoore, whom he didn't know. He introduced himself and said he'd like to talk to him about a news bulletin for children.

'Good idea' said Amoore, 'come and see me at five o'clock.' So he did, and before any discussion could take place Amoore pulled down the blinds in his office 'so the bosses can't see us drinking' and from his cupboard produced a bottle of gin, two glasses and a very small amount of tonic. 'Now speak,' he said, and Edward explained his plan.

'Try it as an experiment for a few weeks and see what happens,' said Amoore. Then he made a remarkable offer. 'I can give you a studio and crew as well as the use of my reporters and correspondents when they have time – and no charge!' It was far more than Edward could possibly have hoped for – full and free access to the BBC's vast newsgathering operation.

'Have you got someone to anchor it?' Amoore then asked. When he heard that young Dimbleby was busy he said: 'You could do worse than John "*Search*" Craven. He interviewed me for that programme on Children's TV and I thought he was good.'

'So, in the space of a few minutes, the world's first news bulletin for children was born,' says Edward. 'Television organizations everywhere have tried to emulate it but nowhere else has there been an editor of television news with the breadth of mind to ensure its success. I doubt if any other head of department in the BBC's history would have shown such foresight, generosity and objectivity. Derrick Amoore was a maverick; he put programmes first and rubbished anything or anyone who stood in their way.'

Edward phoned and offered me the gig. I could hardly believe my luck; to be invited to be the face and voice of a brand-new source of television news with the challenge of winning over a reluctant audience. Plus I did have a security blanket – it would be an on-air pilot for only six weeks and if all went pear-shaped I still had a job on *Points West*. My boss there, Pat Morley, agreed to release me for the twelve days of this remarkable experiment.

Edward would be the editor as, in another huge and significant gesture, Amoore said he didn't want to take control because he knew nothing about broadcasting to children. When I look back, I realize we were so lucky to have Paul Fox and Derrick Amoore in high places –lateral thinkers who could see the potential – as well as Monica Sims, the most powerful woman in television at the time, who as the boss of children's programming was prepared to take the risk.

And it was a risk. There were many adult doubters, including some within the children's department, who feared that a news bulletin in the middle of the 'safe' and almost sacrosanct afternoon programming would be too disturbing, too disruptive, and should not be part of the output. Our mission was to try to win them over as well as the target audience.

I had several meetings with Edward and the big question was: what should we call this new show? We tossed around a lot of ideas and came up with the name *Newsround*. Why? Because it sounded similar to paper round – and older children across the nation were delivering the news through letter boxes, just as I had when I was their age, and younger children aspired to do that. So hopefully the name would connect with our audience in a way that felt personal.

But then Edward said: 'A newsagent puts each child's name against a round, so it's Fred Blogg's paper round or Jane Smith's paper round. So ours will be *John Craven's Newsround*.' He had just returned from a trip to the United States and was fascinated by the American newscasting style which was far more personal. People would say, 'Did you watch Walter Cronkite last night?' not 'Did you watch the CBS News?' 'I wanted to bring that to *Newsround*,' added Edward – and that's how, for two disparate reasons, I got my name in the title. Not many people in the history of television have had that honour.

Picking the right theme tune was also vital. It had to have an urgent, newsy feel to it, be instantly recognizable and be able to be fit any lead story, happy or sad. Mike Beynon, a former film editor who joined our little team from *Blue Peter* as a writer/director, found a winner with a tune called, perhaps appropriately, 'Johnny One Note'.

It was on a disc by, would you believe, one of the top British big bands from the 1950s, the Ted Heath Orchestra. Mike had come across it when he was putting together a *Blue Peter* film about sledging and we used the first sixteen strident bars for us and faded it out before the main tune started. Though its provenance was unlikely, it became an instant hit with viewers and for years to come children would follow me down the streets as though I were a Pied Piper, singing out that tune. Even today people in their forties and fifties sometimes recall 'Johnny One Note' to me, with a happy look of remembrance on their faces.

Traces of it can still be detected in the current, reworked theme – but without our famous bongo drums. And we had a just-as-catchy 'sting' at the end of the show, which Mike pinched from a BBC science programme: didi-dum didi-dum– Didi-da-didy-da!

In our bid to attract Britain's seven- to twelve-year-olds the last thing we wanted them to think was that *Newsround* was just another boring old grown-up news bulletin that had somehow popped up amid their favourite shows. The visual impact was hugely important so we took the crucial decision that I would not sit behind a desk like all the other newsreaders – and like many teachers in those days.

Instead I would perch on a stool in front of a piece of studio furniture that looked a bit like a desk. And I would dress casually, either shirt-sleeved or in jumpers, unlike all the other newsreaders (and, at that time, many teachers as well) in their suits. We reasoned that children were flopping in front of their television sets at home after a hard-working day at school and we didn't want them to think that, by watching *Newsround*, they'd been unwillingly transported back to the classroom.

Edward and I were amused to read quite recently that ITN's Tom Bradby was the first newscaster to have a more relaxed approach and that Channel 5 News pioneered the concept of the newsreader in front of the desk. 'Hang on a minute,' we said, 'what about us?'

Still focussing on the visual impact, *Newsround* would need as many pictures as possible, both still and moving, and attractive maps and captions. As for the journalistic style, we wanted it to be simple without being simplistic or patronizing.

Gone were the days of talking down to the children. It would not be a case of: 'Now sit quietly, children, and Uncle John will tell you something interesting.' We had to look like a proper news bulletin that was just for them. 'I wanted children to feel that our news was very much their news,' says Edward.

Our audience had a thirst for knowledge but hadn't been around on this earth for very long, so they needed explanations as well as information. Too often children were confused by what they were picking up from hectic news summaries on Radio 1 or dramatic headlines in the tabloid press.

Newsround's job would be to unscramble what to them was newspeak gobbledegook. Unlike the main news we would assume no previous knowledge of any subject. Our stories would start at square one and would not hide behind the easy phrases and acronyms used by our senior colleagues (UDI, UN, legislative powers, demilitarized zone, sources close to, etc. etc.). Our watchword from day one was: keep it interesting, keep it simple, keep it short, keep it visual and keep it as positive as realistically possible.

We would tell them stories that interested them – about other children, pocket money, school life, space travel, adventure, the natural world, sport, technology, pop music and so on. Those were the stories they wanted to hear about. But also there were many other, often significant, events happening in the world that we believed they should be told about, in a way they would understand.

Those major events would not necessarily be the lead in the bulletin because we wanted to 'hook' our viewers with stories that would keep them watching, but they would be there. A classic example was the day on which our first report was about a plan to train dolphins to search for the Loch Ness monster – a genuine story, reported by a BBC correspondent, but perfect for us. It was followed by the overthrow of Idi Amin, the Ugandan despot.

And like ITV's *News at Ten*, we always had a funny story at the end, to reassure our young viewers that nice things happened in the world. Years later the satirical puppet show *Spitting Image* had a sketch in which the effigy of its great newscaster Alastair Burnet pleaded with mine for one of my 'end' stories as he didn't have one for *News at Ten*. I think it was, in true *Spitting Image* fashion, about a hamster that ate New York! The *Newsround* team wanted to buy my puppet as a gift for me when I left, but sadly it was too expensive. The production company did agree, though, to lend it for a few days so it could attend my farewell party.

But back to the beginning. *Newsround* was born on 4 April 1972. Edward was the editor and on the writing team Mike Beynon and I were joined by Amanda Theunissen, a colleague of mine from *Points West* who was a first-class journalist and broadcaster. Early that morning Amanda and I travelled from Bristol Parkway to London by train and treated ourselves to a full English breakfast en route at eighty-five pence, not knowing if we would have time for lunch. A wise move, as it turned out, because it would be a hectic first day.

We read all the papers over toast and coffee and picked out a few potential stories. Our little team held an editorial conference in a space allocated to us in a small corner of the foreign news department in the enormous main newsroom on the fourth floor of Television Centre. It was dominated by the clatter of typewriters and telex machines, cigarette smoke and men – hardly any women were involved in the news-gathering process in those days.

We discovered we had a film-editing room of our own and the use of N3, a small, little-used studio. We had very limited access to the graphics team, picture research and the information library and no access at all to news video-editing. Instead we were given an hour from 3.30 p.m. in the main videotape area deep in the basement of Television Centre.

TV news had separate studio teams for each of its two channels, and we were allocated the BBC Two team because, although they had to be on stand-by for any newsflashes, their first scheduled bulletin was not until the early evening. So that was our set-up and we had a format. But what would we put into the bulletin?

The first *Newsround* had something in common with the

first *Children's Newsreel* twenty-one years earlier – an animal story. On *Newsreel* it had been about Brumas, the first polar bear to be born at London Zoo.

Our story was an exclusive – the return of rare ospreys to their nest at Loch Garten in Scotland, with a full background on how these fish-eating birds of prey had been hunted to extinction in some parts of Britain but were now starting to make a comeback.

We had a perfect lead – a story that would grab the attention of our audience and welcome them to this new, newsy experience. And that was how it was to be forever onwards. The most important world event that day, depending on what it was, would be on *Newsround*, but not necessarily as the first story. Our lead would be the 'hook' that drew the viewers in to watch a programme which they knew didn't go on for very long, unlike the grown-up news.

'I was in charge of presentation at the time,' remembers Edward, 'and the announcer always had to say that in five minutes' time it would be *Blue Peter* or whatever, but first here's *John Craven's Newsround*. In those early days, it was a great advantage being sandwiched between two programmes they wanted to watch.' And before remote control buttons were invented it was quite an effort to cross the room and switch channels for just a few minutes, which also helped.

On that first day, most things went surprisingly to plan. Sadly, no recordings exist of the early editions of *Newsround* and I can't remember any of the stories from that first edition apart from the ospreys. It was a hectic seven hours and shortly before we were due on air I found myself lost along the circular corridors of Television Centre and in full panic mode, trying to find my way to the studio from the video-editing area in the bowels of the building. Would this brand-new programme have to be aborted because they couldn't find the presenter?

Luckily, I asked directions from someone who, bless him, actually led me through this unfamiliar maze to the studio. I made it in time to calm down, have my nose powdered by the make-up artist and plug in the earpiece that connected me to the control gallery.

I heard the director, a seasoned veteran of many dramatic moments in news studios, say: 'Two minutes to go and I still don't have some of the scripts. So everyone repeat after me: "Our Father, which art in Heaven . . ."'

That broke the tension and He must have been listening because the missing scripts arrived with seconds to spare, the programme started and ended on time (very important with live television) and everything in between seemed to go well. Edward leapt from his editor's chair, gave hugs all round and took the team for a celebratory drink in the BBC Club. Our dream was coming true – but would it last?

Every Tuesday and Thursday for the next six weeks Amanda and I took the 7.50 a.m. train up to London and became such regulars that Gordon, the chief steward, had our coffee already poured in the dining car when we boarded at Bristol Parkway. Our regular table companions, who had joined the train earlier and made sure we had seats, were Professor Leonard Hirsh of the Royal College of Music (as a sideline he played the gypsy violin for Marilyn Monroe and Laurence Olivier in a scene from the film *The Prince and the Showgirl*) and Claude Adkins, who was a specialist in oriental art.

Over the bacon and egg or kippers until Swindon we put the world to rights and then these two learned gentleman helped us scour the newspapers for stories than might be useful for *Newsround.* They loved being volunteer researchers for a programme they would never see, and picked some good ones. It was an intellectually charming calm before the storm that was awaiting in the studio.

On the second day I bumped into Derrick Amoore in the corridor and he said, 'Well done,' and I don't recall him taking any further interest in us. But what about our audience – what were they thinking about this new kid on their block? Because we didn't really have access to much audience research in those early days, we could only rely on word of mouth.

So I was thrilled when a friend told me that he was in the garden when his eight-year-old son came rushing out to tell him about a sunken treasure ship that had been found. 'Where did you hear about this?' asked my friend. 'On the news,' said his son. 'Don't be silly,' replied his dad, 'the news isn't on until a quarter to six.' Then came the reply which convinced me we were hitting home. 'I heard it on MY news!' If one small child was reacting like that, hopefully so were millions more.

Much to my surprise my jumpers started to make news themselves, with angry letters to the quality newspapers about falling standards. How dare someone on the BBC read the news without a jacket? Well, I dared to, and it worked, and I built up quite a collection of woolly pullovers, kipper ties and voile shirts with large collars – all of which were my personal choice because there was no wardrobe department to advise and no clothing allowance.

But there was a price to pay for looking casual.

Because I was on a stool in front of a desk there was nowhere to put my scripts so I had to rest them on my knees.

On top of all my other concerns, I had to keep a tight grip on them to make sure they didn't fall to the floor. In my earpiece one afternoon I heard Edward's voice from the editor's chair in the control gallery saying we were dropping one of the stories because the programme was overrunning. So I let that script slip into the bin by my feet only for Edward to then say they'd got the calculations wrong and the story was back in. The previous film report was just coming to an end so viewers saw me lurching out of shot to retrieve the script. 'Won't be a moment,' I said, 'I've just lost something,' and then on we went.

Slowly the team began to fine-tune *Newsround*. We were moved from foreign news (where, to be honest, our presence had not been welcomed) to a windowless room near the main newsroom and just next to our studio, N3. Amanda says she found many of the news staff unhelpful, even though everything had been officially agreed:

> We had to beg for time in the newspaper cuttings library, at the photocopier, everything. No one thought the programme would last and they couldn't be bothered with it or us. I nearly caused a strike in the studio because I moved a globe we needed for a story about six inches and had crossed the demarcation line. It was the first time I had worked anywhere in the BBC apart from friendly *Points West* and I couldn't get over the hierarchy and the formality.

The *Radio Times* later reported that 'at the beginning the hardened, rather cynical journalists of the news division regarded it with barely disguised condescension and suspicion'. Mike Beynon, too, remembers that although some people in the newsroom, including my old friend from *Look*

North in Newcastle, John Exelby, were really happy to work with us and show us the ropes, there was also an undercurrent of hostility:

> Many people saw us as just adding to their workload without getting any extra money for it. But the boys in news graphics were particularly helpful because we set them some interesting tasks such as animated maps [which involved pulling out strips of cardboard to reveal what was underneath] and cut-out animals.

John Exelby was then the BBC's news organizer – a classic example of the corporation's strange affection for weird job titles. Because of its unpredictable nature, how can news possibly be organized? Manipulated sometimes, it could be argued, but not organized. Some members of staff disguise their job description by using its initials, but that can also have unintentional side-effects. It's said that one engineer used to answer his phone with his acronym 'E-I-E-I-O', just like Old Macdonald on his farm.

Anyway, John recalls the reaction to the arrival of *Newsround*:

> While many people in the hard-nosed newsroom welcomed what they regarded as an exciting development which would attract a new audience to TV news, others were not so helpful. I'm not aware of any obstruction but more a feeling that it couldn't work because it was not possible to simplify stories enough for young people without distorting their meaning.
>
> Exactly the same thing happened fourteen years later when I suggested and then edited the three-minute bul-

letins which were at the heart of the BBC's first daytime schedule. Sceptics, including people who went on to hold the most senior positions in TV News and Current Affairs, said it wasn't possible to fit six stories with pictures and interviews into three minutes. But, as with *Newsround*, they were wrong. They had overlooked that the skill lies in the discipline of the writing and the editing.

At the end of our six weeks the 'on-air pilot' *Newsround* was deemed a success – we had attracted about half the children of Britain to their very own news service and Edward announced that we would be back with a six-month run from September to March, but still only twice a week.

'We spent our days without seeing daylight,' says Amanda. 'Writing for *Newsround* was hard work but excellent training. Thirty seconds was the usual length for a story, forty-five if it was really important, and having to do it in that specific time really honed my storytelling skills in a way that stood me in very good stead. You had to get straight to the core and waste no words, and that's how I teach people to write scripts even now.'

How true that was, and still is. Writing a good, interesting and comprehensible script for *Newsround* was a constant and fulfilling journalistic challenge, especially when, as often happened, there were complex issues involved and there wasn't much time to get things right. I'd sit at my battered old typewriter (later to be replaced by an Amstrad computer) and gather all I could about the story I was handling: still and moving pictures, news clippings, telex reports – and then say to myself: 'I am eight or nine years old. How on earth am I going to understand this?'

Then I'd spend a few minutes thinking it through, sorting out the details in my mind (which facts to leave out are just as important as which to leave in when you have perhaps 200 words at your disposal) and then hit the keyboard.

We would try to find a positive spin on any disaster, without straining credibility. An often-quoted example is when an airliner crashed on take-off in Nairobi. The evening newspaper headline was 'Airliner Crash Horror'. On *Newsround* the headline was 'An airliner crashes on take-off – but ninety-three people walk away from the wreckage.'

It was so important to clearly understand the story before you passed it on to others, especially the young. The last thing we wanted was to be accused of brainwashing children (and to my knowledge we never were), so at all times our treatment must be fair and balanced without being boring. It was a commandment that made me a much better reporter.

When I explained the *Newsround* philosophy to a Greek broadcaster at a Eurovision conference, he raised his eyebrows and cautioned me about the long-lasting effect we would have. 'Always bear in mind,' he said, 'that there is an old saying which goes: "This is what I believe because they told me it was true when I was young."' It was a warning I never forgot.

Some critics accused *Newsround* of destroying the garden of childhood. They thought children should be cocooned from the outside world, conveniently forgetting six- to twelve-year-olds see newspapers, pick up playground gossip and catch part of an adult news bulletin before they flee the room. In this modern era of communication, no one can escape the news machine.

My response was that, far from destroying the garden of childhood, we were in there with them. *Newsround* provided a ladder so that our young viewers could climb the wall of that garden and peer over the top into that outside world. I was there, as a familiar and hopefully trusted adult, holding the ladder and explaining to them what they were seeing so there would be no misunderstanding.

We rarely tackled political or industrial issues, unless they affected children in some way. But there was a long-running industrial dispute in the Midlands and families of strikers were starting to feel the pinch. So we decided to run a story about it and I was as even-handed as I could possibly be, explaining the issues on both sides.

Later a union official telephoned to thank us for being so fair and accurate. *Oh dear*, I thought, *the other side must now be fuming!* But no, a little while later we were praised by the management for our coverage. Then I thought: *If both sides agree on our version of what's happening, how come they don't get around a table?* Which they finally did.

Our little team stayed the same (including our unpaid, overqualified helpers on the train) and gradually we began to be more accepted in the main newsroom, although we still had limited access to the mighty facilities of BBC News. We still had to beg a lot but I kept reminding myself that we were being given these resources totally free.

Thursday 30 November 1972 was a big day for me because for the first time I was not there at five o'clock when *Newsround* went on air. Instead I was back in Bristol becoming a father.

Marilyn had gone into labour the night before and when Emma came into this world I wanted to be there. She took her time and I paced the waiting room all day with far more important things on my mind than what would be the lead item on that evening's programme.

But come five o'clock, and with still no news from the delivery room, I couldn't resist tracking down a TV set and finding out how they were coping without me. Very well indeed was the answer. There was Richard Whitmore, one of the stars of the grown-up news, sitting in for me and the first thing he said was: 'I know I'm not John Craven but he's rather busy today,' and he went on to explain why there would soon be another member of the audience. Then it was on with the show as usual.

Emma finally arrived that evening, safe and sound, and I was there for her birth – a wonderful experience that I will never forget, a deeply personal moment that really binds two people together, even though we did have a little argument over whether I was giving Marilyn enough gas and air. Later, with both my girls safe and sound, I headed home and began to think how different life would be now that we were a family of three.

Being a father made me realize what a debt I owed to the parents of Britain, who were allowing me to tell their children what was happening in the world, often when they themselves were not in the room. They were letting me act *in loco parentis* and I vowed that I would never let them, or their children, down – and I hope I never did. My other daughter, Victoria, was a little more considerate in timing her arrival – she chose a Saturday afternoon.

What gave our team a tremendous boost from day one was the support of many of the BBC's best news reporters and correspondents – the likes of John Humphrys, Martin Bell, Keith Graves, Michael Buerk, Brian Barron, David Willey and Tony Lawrence. Reg Turnill, the much-respected aviation correspondent, became *Newsround*'s space editor when our original one, the wonderful Patrick Moore, became too busy to help us.

Patrick's enthusiasm helped win over the nation's children to the wonder and excitement of space exploration but I had the unenviable task of asking the great man if he could possibly speak a little more slowly. The audience, we learned from their letters, was having difficulty in following everything he said. Bless him, he did try his very best but attempting to stem his avalanche of words was nigh-on impossible.

I remember ringing him one day at his home at Selsey in West Sussex about a potential story and his mother answered the phone. 'Could I possibly have a word with Patrick, Mrs Moore?' I asked. 'Just a minute, John, and I will see if I can find him.' There was a long delay before she came back to the phone. 'He must have gone out because his bicycle clips aren't hanging up.' This formidable man, who thrilled the country with his tales of distant planets and rockets hurtling through space, was pedalling around Selsey on his bike.

One day my mum spotted him at Leeds railway station and approached to say hello. Patrick could not have been nicer so she plucked up courage to say that the night before he had been on her son's programme – and he told her how much he enjoyed working on the show. How typical of Patrick and how fortunate for us that he, like those distinguished newsmen, was willing to support the upstart *Newsround*.

They sent us special reports from wherever they were in the world, signing off with 'for *John Craven's Newsround*' and making the effort to use words and references that our viewers would understand – and that really gave me a kick.

The war in Vietnam was at its height in our early days and I wanted our audience to know what it must be like for children caught up in the conflict. Martin Bell was one of the BBC's greatest correspondents, at his brilliant best in war zones, and because of his reputation it was with some trepidation that we asked him if, should he have the time, he could report for us on the plight of Vietnam's youngest people.

We shouldn't have worried because he was certainly up for it and filmed for us in a village that had been destroyed by fighting. When I watched it coming in on the satellite I felt so proud that one of the BBC's top men had crafted a moving, pitch-perfect dispatch carefully worded and filmed for *Newsround.*

Martin introduced our audience to Nuen, an eight-year-old who had lived through it all and was making a new start with his family. It was our main story of the day but it was also prominently shown, unchanged, on the *Nine O'Clock News* (much to Martin's surprise, I suspect, because he had filmed it specifically for us). Another of his reports was from an orphanage where red tape was stopping children whose parents had died in the war from being adopted.

'War is about people and one of the features of modern warfare is that children are so much caught up in it because they are unable to get out of the way,' Martin said later, explaining that he would try to relate their experiences in a way the young audience back home would understand. 'I think writing for *Newsround* made me a better journalist. It taught me to simplify and try to avoid jargon, which was a very good lesson.' And he told the *Radio Times*: 'I always say you won't get a more intelligent and perceptive audience anywhere. Sometimes if I've had a quiet time in the Middle East I've taken off to the Golan Heights and done a special piece for them. Of course, there are times when the pressures of the day-to-day news are too great.'

Keith Graves told me that when he returned home from a long foreign assignment, his children rushed to greet him with the words: 'Dad, you were on *Newsround*!' He ruefully reflected: 'In their eyes I had finally made it.' That, I suspect, is why many of them were willing to put in that extra effort to write a script for us – it was the only time their young families ever got to see what Dad did for a living.

And Alan Protheroe, who was deputy editor of television news, told Edward:

> What delighted us was the way it quickly became clear you were going to carry real news. I began to take you seriously when you somehow got on to the end of a satellite booking bringing in pictures of the Turkish earthquake and demolished the lead for the early evening news! But our philosophy is to get the news on the air at the first opportunity so we didn't mind at all.

Well, some of his team did!

When I had done my reporting course some years before I had watched in admiration as the BBC's great newsreaders (never newscasters – that was an ITV innovation) Robert Dougall, Richard Baker, Kenneth Kendall, Michael Aspel and Bob Langley effortlessly informed their audience. All hell might be breaking out in the newsroom, the control gallery and the world but you would never guess that from their smooth performances.

Late scripts would be handed to them by a floor manager and they would read them sight unseen. Major stories would be dropped to make way for even more important breaking news. Changes would be made on the upcoming section of the autocue while they had their head down reading from a script. They sailed through it all without raising an eyebrow although, as I was to discover, the heart rate certainly goes up when there is only you between an excited news team and a few million people watching at home.

In those days, autocue was typed onto a long strip of paper and fed through a machine which projected the words onto a clear screen in front of the camera. Newsreaders could make little alterations to the words to suit their style. I once caught sight of Robert Dougall's autocue and he had written notes to himself like 'long pause here' and 'slight smile'. Now I was the junior member of that small, illustrious club.

It's easy to forget that, during our early years, *Newsround* was the first BBC television news bulletin of the day and I ceased to be surprised when staff in pubs (shock revelation – children's TV presenter drinks beer) talked to me about stories they had seen on the programme. For shift workers, and even for some of the showbiz stars I got to know in the BBC Club, like Les Dawson, it was the only television news they had time to see and they took us seriously.

As we approached our 2,500th edition in 1985, the television critic Mark Lawson wrote in *The Times*:

> *Newsround* is one of the glorious successes of British television. Its story is one of initial opposition – early controversies ranged from the advisability of showing the Six Day War to

the correctness of a BBC presenter wearing sweaters – defeated to become an institution.

At a time of great debate about what children should be shown on television screens the *Newsround* rules of conduct deserve examination.

First, there must be no lingering emphasis on death, no close-up carnage: the Colombian earthquake was depicted, with little loss of impact, through the broken limbs of buildings littering the streets. Murders and rapes are not mentioned although the capture of the 'Yorkshire Ripper' was. The children's equivalent of industry – school – is well covered.

The great strength of *Newsround*, and one that presumably attracts many adults among its audience, is its necessary mission to make the intricate simple. It is hard to see any remaining credibility in the argument that children are being told and shown things they need not know.

John Craven is the only newsreader in Britain to get his name above the title and he is the vital link between young viewers and the news. In a smile and a sweater, he fits brilliantly between 'Hello again' and 'Bub-bye' a mix of global disaster and dancing dolphins. He has made the role his own and it is impossible to imagine that any other presenter could do it better.

It is a nice irony, and one in which *Newsround* should take pleasure, that while the rest of the BBC News output has spent a decade shuffling desks and presenters in an attempt to find consistent intelligent authority, the originally unwanted little brother got the format right first time and maintains its level of excellence.

Thanks, Mark – all very positive and typical of many press reviews over the years, and can I just add that the reason we mentioned the arrest of the Yorkshire Ripper was that we knew there had been much playground talk about his murders and we thought that reporting his capture would reassure our audience. Putting a closure on disturbing events is something we have done many times.

But where should we draw the limit with stories that might be perplexing to children? One test came towards the end of our second six-month run, in March 1974. An attempt was made to kidnap Princess Anne while she and Mark Phillips were being driven back to Buckingham Palace after attending a film premiere. Just a couple of hundred yards from the Palace, their Rolls-Royce was forced to a halt and a gunman called Ian Ball tried to eject the princess from the car.

Four men, including her detective and chauffer, were shot trying to defend her before Ball ran away. It was all hugely dramatic – especially when a ransom note addressed to the Queen demanding £3 million was found in his car. Fortunately the royal couple were unharmed, the wounds were not fatal and Ball was arrested. He was later sentenced to life imprisonment and placed in a psychiatric institution.

This happened on a Wednesday night and the foiled kidnap was splashed across newspapers around the world the next morning, so my natural instinct was to do a follow-up on the Thursday *Newsround*. Edward was filming abroad and Monica Sims was keeping watch over the bulletin. She needed convincing. Would our viewers be troubled if *Newsround* ran the story because Princess Anne had recently co-presented a *Blue Peter Special Assignment* programme with Valerie Singleton? Was it a story too close to home? Might some children fear that Valerie – much loved by the audience – could be the next victim? An interesting point.

I said we would reassure viewers that Princess Anne was safe and had been very brave, that the attacker had been caught and posed no further threat and that security would be stepped up around the royal family to make sure nothing like it would ever happen again. Plus it would be the talk of the playgrounds, especially after her *Blue Peter* appearance, and many viewers might find it odd if we ignored it. Monica agreed, and it was our main story that night.

A week later, at the end of our run, Monica and Edward made a big decision. When we returned on 9 September 1974 it would be for four days a week, not two, and we would transmit throughout the year except for a long summer break. The baby of television news was now standing and walking.

8. *Newsround* Gets Bigger

Newsround on four days a week meant I couldn't travel daily from south Gloucestershire to London as it would be too tiring, so on Monday and Wednesday nights I stayed in a small commercial hotel near Television Centre. It was not a satisfactory arrangement, especially with a toddler and another baby on the way. So after daughter number two, Victoria, was born, Marilyn and I decided to move to London.

We said goodbye to leafy country lanes in south Gloucestershire and bought a small Victorian semi just off Goldhawk Road in West London. The only rural connection was in the name of the suburb – Shepherd's Bush – but we loved the house and for me the bonus was that it took fifteen minutes to walk to work rather than a commute of nearly three hours.

In the office, there were changes. Amanda had already opted out, so only Mike Beynon and I remained from the original team, which more than doubled in size to cope with the extra work. We needed a bigger office after being confined to our small, windowless one and I came very near to demanding in my new contract that I work in the daylight.

In fact, we got more than just one window – we got dozens of them. The walls of our new office were made entirely of glass. It was on the top floor of Television Centre and had been designed as an observation suite where BBC bosses could entertain important visitors and take them on to the balcony to get a good view of London.

But those days were over and now it was a corridor-like office; one half was ours and the other was taken over by Barry Norman and his team from *Film 74*. Because he was such a nice man, we let Barry take a short cut through our office and he always stopped for a chat. Though we were now working in brighter surroundings, the myriad windows had their drawbacks. They were not double glazed, so the place was cold in winter and like a greenhouse in summer, but we did have a great view.

Edward Barnes, who still supervised the show in addition to all his other work in the children's department, appointed Jill Roach as the editor. I had worked with Jill on *Search* and knew she was perfect for the job. She was young, decisive and talented and some of the old hands in the main newsroom didn't know what had hit them when she came calling for editorial favours. Jill was the only female editor in television news and, as she recalls:

> Some of the newsroom staff were reasonably enough suspicious of this small band of outsiders, some non-journalists, whom they'd been told to support but did not control. It was potentially a delicate situation.
>
> Developing good relations and gaining the newsroom's respect were vital to *Newsround* getting off the ground and staying airborne. Professionalism and charm were the order of the day. At the heart of news planning was the weekly news editors' meeting, where known upcoming events were laid out and coverage discussed. I, the *Newsround* editor, was 'admitted'. There were usually around eighteen of us,

seventeen men and myself, the only woman in the group and definitely an outsider and something of a novelty with no power, only persuasion.

This group, the editors of the main BBC news bulletins, were all courteous as long, that is, as *Newsround* didn't get in the way of their stories. I didn't feel overt sexism from them. In fact I was so completely absorbed in the shaping of our programme, in establishing how and what *Newsround* could leverage from News without falling out with anyone, that I was gender blind myself. I just asked for what we needed.

I was, however, reminded that *Newsround*, or indeed I, didn't always have the respect that we felt we had earned. Some seven years on, and several British Academy Awards later, a news executive said to me that if I wanted to be a deputy (not even full) editor in the actual news department they would look seriously at my application. I am not sure if this was sexism or because *Newsround* was still considered 'only a children's programme'. Either way I wasn't impressed.

Jill steered our ship brilliantly and consistently impressive viewing figures proved that the decision to add the extra days had been a good one – 5.5 million people were watching, including 55 per cent of all five to seven-year-olds. Back at home, Marilyn continued to be my manager/agent as well as my best critic. 'How did the show go down in Thornbury?' Edward used to say when we lived in Gloucestershire.

Now it was 'What did Shepherd's Bush think of the show?'

One day I phoned Marilyn from the office to tell her about some big new development on the programme and she said: 'Sorry, I can't talk right now – we've lost pink rabbit.' That put everything into perspective – the search for a much-loved and missing soft toy was top priority for the Cravens.

On another occasion, when I must have looked disappointed because Marilyn and the girls had missed what I thought had been a particularly good bulletin, Mrs C pointed out an important truth. 'If you were driving a bus around Shepherd's Bush Green,' she said, 'you wouldn't expect me to stand on the corner and say: "John you made a great left turn."' We had a good laugh and reactions like that helped me keep my feet on the ground – I was just a dad who went out to work.

When Emma and Victoria were small we took our holidays in places where we could be guaranteed a little privacy – though public recognition came with my job, it wasn't fair on the girls if their dad was constantly being asked for autographs. For some time we spent a couple of weeks each year in a cottage partly owned by my old friends from Harrogate days Peter and Susan Rose, in an idyllic, rustic corner of southern France.

We would visit the big municipal pool in the nearby town and, of course, no one had the faintest idea who I was. Until, that is, French television began to show episodes of *Newsround* to help their young viewers improve their English. I did get some letters pleading with me to speak *plus lentement* (a little slower) but didn't really think about the impact until that summer. When we arrived at the pool a group of French children rushed up calling out 'Jean Cra-von'. No more anonymity in Bergerac, but it didn't really spoil our holidays.

I have never been part of the show business side of television, though occasionally Marilyn and I accept invitations to glitzy events. One night we went with our friends from children's television Keith Chegwin and Maggie Philbin to a gala evening at a Park Lane hotel.

We'd forgotten the invitations and when we arrived in the pre-dinner drinks room it was deserted apart from empty glasses. 'I didn't realize we were this late,' said Keith. When we walked into the dining room everyone was already eating, and as Keith and I searched for our table people were greeting us and shaking our hands. Then we were asked 'Are you guests of the bride or the groom?' and the penny dropped – we had gate-crashed a rather grand wedding reception and should have been at a different hotel altogether.

There is one sure way to get to the right place at the right time, and that is to go in a police car. I'd been invited by the Metropolitan Police to present safety awards, but after the event had been publicized I realized that was also the evening of the BAFTAs, and *Newsround* had been nominated.

When I explained my problem to the police, I was told: 'Don't worry. Our awards are in the early evening and we'll make sure you get to the BAFTAs in time.' And they did, in one of their cars with, at times, the blue lights flashing. Very exciting, and to top the evening off *Newsround* won a BAFTA for best children's factual programme.

Over the years the framework of a typical *Newsround* day stayed much the same. The team arrived at around nine and checked through all the newspapers for suitable stories which we could follow up. They'd also sift through the news agency tapes and press releases and confirm the arrangements we had made for covering stories already in the diary. Jill joined us after attending the daily conference of the main news editors which mapped out their agenda for the day – what the main stories would be, where the correspondents and reporters were being sent and so on. She'd have noted everything that might be useful for us.

Then we drew up a longish list of our potential 'ingredients' and talked them through at our 10 a.m. meeting. It was always a lively affair, with lots of opinions bouncing around, and afterwards a provisional running order was compiled. After they'd been researched, some of the stories got rejected for many different reasons and others were given the go-ahead. Just how many we pursued would depend on the length of bulletin, which varied day to day – anything between six to nine minutes.

Each team member was handed one or two stories and set about putting them together in our own special way while our reporter was dispatched with a film crew to compile the day's '*Newsround* exclusive'. As the morning progressed there was lots of 'phone bashing' plus trips to the foreign news desk – 'Can Michael Buerk in South Africa write a version just for us?' – and questions for the home news organizers: 'What time do you expect the film of the Queen's walkabout?' Another vital port of call was to the news information department, which had a massive collection of press cuttings on every subject under the sun. Slowly, another *Newsround* was beginning to take shape.

Things had often hotted up by the time we held our second meeting at 2 p.m., which was also attended by the studio director, when Jill outlined the running order of the bulletin. Big stories might have happened during the morning and needed to be added and, of course, others might break right up to the moment we went on air – and often did. Each story was given a specific number of seconds. The writer could plead for a few more if it was really justified but that would mean that same amount would have to be chopped from another story.

My role during the day was to put together one of the big stories, maybe keep an eye on some of the others, and often write our 'and finally' – the funny story at the end. Films were being put together in our cutting room and at 3.30 p.m. we got access for just one hour to videotape editing. During that time, any reports we had requested from regional newsrooms would be sent 'down the line' and we could also tap into satellite transmissions coming in from BBC correspondents around the globe.

These sometimes included reports done specially for *Newsround*. One of my fondest memories is of John Humphrys sending a dispatch for the main news from China. When the video editor was satisfied it had been recorded successfully John was told the line was closing (satellites were incredibly expensive) but he insisted: 'You can't cut me off yet. I haven't done my piece for *Newsround*'. The line stayed open for another couple of minutes and we got our report. Thanks, John!

Back in the office, Jill was putting final touches to scripts and adding up the total timings to make sure the programme didn't overrun. With a few minutes to go to transmission I made my way to News Studio 2 (we had outgrown our original and tiny Studio 3) where, in the small dressing room, a make-up artist was waiting to prepare me for the camera! It didn't take long, but for me this was the moment of calm before the controlled tension of a live news bulletin where anything can happen.

Then the floor manager would hand me the scripts, I would check through the wording on the autocue with the operator and put in the earpiece that connected me to the studio. I always had what was called 'open talkback', which meant I could hear not just the director and the seconds being counted down but everything else in the control gallery. I found it reassuring, and if anything was going wrong, such as a problem with the videotape machine, I was well aware of it. On the desk behind me was a trimphone (remember those?) which wasn't connected. It was just a 'prop' but if anything went wrong I would pick it up and pretend to be using it while actually getting instructions through my earpiece.

Though *Newsround* was a complicated programme, both editorially and technically, our whole purpose was to make it look as smooth and seamless as possible for our young audience. There were some unexpected moments, such as the night one of the robotic cameras went bananas and I seemed to twirl down to the studio floor from the ceiling, but on most bulletins everything went as planned.

If there was a late developing story, room had to be found for it by dropping or by cutting short – sometimes on air – one of the other stories, and that was when the adrenaline really started to flow. I would get an indication in my earpiece that something was happening while I was in the middle of another item or, if I was in vision at the time, I'd pick up my trimphone. The new script would be slipped to me by the floor manager and the director would tell me when to read it and whether any pictures or maps would be shown on the screen. How could anyone then doubt that *Newsround* was a 'real' news bulletin?

While I was reading the 'and finally' I could hear the production secretary in the gallery counting down the seconds to the moment we had to be 'off air' and if I could time my words so that we ended as she said 'zero' it was, for me, like hitting the jackpot! Another *Newsround* had been and gone and hopefully the nation's children were a little better informed.

In 1976 Lucy Mathen joined the team as our first full-time reporter and she was also the first British Asian woman to report on the network. Lucy had learned her journalism on a newspaper in Croydon and quickly adapted to the special style needed at *Newsround*. Lucy became a much-loved personality and we worked well together. Among other things she co-presented our special series *Newsround Africa*, which told the continent's story from earliest times, through the colonial era and slavery, to the present day.

People still ask me about her and are surprised to hear that she gave up television to train as a doctor. She became an eye specialist and then set up a charity called Second Sight, which does amazing work in India's poorest state, Bihar. Second Sight's simple yet challenging aim is to cure Bihar's people of treatable blindness through cataract operations.

Surgeons from Britain spend their 'holidays' there carrying out operations and the charity trains Biharis in the skills needed for their vital work. Well over 380,000 people so far have had their sight restored and I am honoured to be a patron of Second Sight and to salute Lucy for her inspiring work.

As our reporter, Lucy was succeeded by Paul McDowell from Northern Ireland, who quickly became a favourite with viewers and who stood in for me on occasions. Paul was a first-class reporter as well as being a great team member and *Newsround* really missed him when he moved to other programmes.

Many other talented people worked on our production team in that multi-glassed office, including Bea Ballard, whom I first met when she researched my weekly children's column in the *Radio Times* and who was to become Michael Parkinson's executive producer, and Helen Fielding, creator of Bridget Jones. From her home in Los Angeles Helen recalls those days:

> I have very fond memories of my time on *John Craven's Newsround*. I was in my early twenties and not the most together of journalists. A lot of Bridget Jones' experiences were based on my experiences there – though John Craven was much more understanding and professional than Bridget's boss Richard Finch. *Newsround* was actually a very difficult news programme to do. The reports were very short, and because the audience were children, you couldn't assume that they understood the background to anything. It made me realize how fuzzy all of our understanding of these situations really is. John was a master at the succinct global précis, and could always get us out of scriptwriting peril at the eleventh hour.
>
> At that point there was still some idea that I might make it as an on-camera news reporter. I had good ideas for stories, and I think I was good at writing, but timekeeping, and keeping my mind on the task in hand have never been my strong points, and I think I must have given them a lot of near heart attacks, but at the same time given myself some good material when Bridget Jones came into being.
>
> The whole scene in the *Bridget Jones' Diary* and movie where she's supposed to be sliding down the fireman's pole on camera and messes the whole thing up came from an

incident on *Newsround*. When Princess Diana was giving birth to Prince William I was, excitingly, given the job of reporting live from outside the hospital where the whole thing was taking place.

It was quite a high-stress situation. There were crowds watching on the other side of the street, and all the news reporters were outside the hospital behind a barrier, taking it in turns to talk to camera when their shows went on air. When it was my turn I was, naturally, very nervous and preoccupied with whether I looked nice. The person who was supposed to cue me forgot to, so I was live on air for quite a long time, fiddling vaguely with my hair, when someone suddenly yelled, 'Helen! Go, go, go, go!' I panicked, took a deep breath then announced, 'Errr, the baby hasn't been born yet but it's all very exciting. Now – back to the studio!'

I think that was the last time I was given a go at presenting, but there were previous – only slightly more successful – attempts: one reporting on Choirboy of the Year when I just looked very startled, and another when I was on a bicycle reporting on London traffic and looked as though I was about to fall off.

It was in the days before digital, or even video, when news stories were shot and edited on film. We had to cut our stories – which were always very short – and then, when the clip was ready, physically take the reel of film to the machine which transmitted it live. I was always, as I still am, a bit of a perfectionist with the detail in stories, so I was never really capable of finishing on time. There were several occasions when John would announce 'And now, a report from Helen Fielding,' and I was still either cutting the said report, or running frantically with the reel of film from the editing room to the transmission machine, so that John would have to say, 'Well, we seem to be having some technical problems with that report from Helen,' and move on to something else.

But John was always very calm and gracious about mistakes. And one of the nicest things was that those were the days when the BBC bar still existed. We all used to repair to the bar afterwards to decompress and laugh about everything that had gone wrong over a drink. I think that was what gave me the idea in Bridget of the friends sitting round with a glass of wine, making everything all right again by having a laugh and supporting each other.

They were happy times and I'm very grateful for everything that I learned from *John Craven's Newsround* and all the great material it gave me.

Another distinguished *Newsround*er was Nick Robinson, later the BBC's political editor and now a presenter on Radio 4's *Today* programme. In his book *Live From Downing Street*, Nick wrote:

> As a child I had my own window on the world in the shape of the first ever children's new bulletin, *John Craven's Newsround*. Its presenter and inspiration, John Craven, who, in my memory at least, always sported a bright yellow v-neck jumper over his shirt and tie held the hands of my generation as he took us on a tour of what was happening around the globe. He did it in a way that was simple. Engaging but never patronising.

Years later, as a graduate trainee at the BBC Nick worked on the show that had first got him interested in the news:

Writing short scripts that a seven-year-old would understand was perhaps the toughest journalistic challenge I've ever faced.

One day in October 1987 I spent many hours writing and rewriting the story of what became known as Black Monday, trying to explain not just the stock market crash but what the stock market was. The Middle East process was equally testing. When I was asked to interview a man about tortoises in the *Blue Peter* garden I was grateful for some light relief.

In August of that year news emerged of a horrific massacre in the little town of Hungerford in Berkshire, where a man called Michael Ryan had gunned down sixteen people, including his own mother, and injured another fifteen before turning the gun on himself.

These days such dramatic, and thankfully rare, news would immediately be spread on social media or the 24-hour news channels, but back then the BBC had the straight choice between breaking it on *Newsround* – the next scheduled bulletin – or asking the TV newsroom to produce a special newsflash. Nick remembers:

> I watched in awe as John persuaded the BBC bosses that he should be trusted with covering the massacre rather than allowing the 'grown-up' news to take over his slot. After all, he argued, thousands of children would be watching in any case, having tuned in to see *Newsround*, and he knew how to tell them the story in a way that would limit their distress.
>
> In the general election that year *Newsround* secured a larger audience than the specially extended *Nine O'Clock*

News. It's just possible that viewers found its storytelling easier to understand and, frankly, less dull. It was a lesson I've never forgotten: not that you should treat everyone like children, but that if you don't engage people, whatever their age, or you assume too much knowledge, they will quickly switch off.

Nick was referring to our mock elections, which, since 1983, we had been running alongside the real thing with the idea of explaining the electoral process to our audience and involving them in it. Hundreds of schools took part by nominating candidates, holding hustings and safeguarding ballot boxes. Hundreds of thousands of children voted and we reported the results, with the Hansard Society overseeing the voting to make sure all was in order. Oddly enough, the results tended to be a little more conservative than the real general election.

Because good stories about the animal world were a number one priority on *Newsround* we wanted to know why the World Wildlife Fund had decided to change its name, and a good person to ask was its president, the Duke of Edinburgh. He agreed to be interviewed in an office at Windsor Castle but unfortunately we could hardly hear what he was saying because a military band was playing outside the window.

The Duke, who obviously is very media savvy and who was on a tight schedule, saw the predicament facing our sound recordist but said he could not ask the band to stop playing because bands had been doing this at Windsor every day for nearly a thousand years.

'But,' he said to one of his entourage, 'we can ask them to play a little quieter.' Off went the man, and a couple of minutes later the noise level dropped and we could start the interview. That, I thought, was real power – to be able to command a military band to turn down its volume! Incidentally, the WWF name change was because many languages do not have the word 'wildlife', so the organization became the World Wide Fund for Nature, but still kept its original initials.

One of WWF's keen young investigators was Mark Carwardine, who exposed on *Newsround* the cruel way in which young chimpanzees were being exploited in Spanish holiday resorts. Mark had discovered that unscrupulous gangs of beach photographers imported them illegally from Africa, often after their mothers had been shot by hunters.

Shortly before we started filming, Mark had been beaten up as he was trailing one of these gangs, so determined was he to collect evidence. They drugged the chimps, took out their teeth so they could not bite tourists and dressed them in children's clothes. Then they touted them round beach bars, restaurants and nightclubs, dropping the helpless little creatures into the laps of holidaymakers, taking photos and then demanding payment. When the chimps got too big to handle, they were killed. A nasty business indeed.

Mark fearlessly reported this racket for *Newsround* and appealed to children and parents going on holiday to Spain not to have anything to do with the abhorrent practice. It must have worked because in following years there were far fewer of them on the beaches.

After leaving the WWF Mark became an eminent wildlife photographer and a world expert on whales and dolphins, writing many books about them. He is also one of my best friends and recently I joined him on a holiday expedition he was leading to the Baja California peninsula in Mexico, where we saw blue whales, humpbacks and dozens of grey whales with their young. One of the 15ft 'babies' bobbed up next to our small boat and I rubbed its nose. Not many years before, other men in boats had gone there to kill them but we went in peace for what many described as a life-changing experience. Of all my memorable wildlife encounters this was one of the best.

Through the seventies and eighties the Troubles in Northern Ireland dominated our news agenda and we were mindful that many of the children who were deeply affected by them were also part of our audience. We did a special *Newsround* edition from the province with children from both sides of the religious and political divide, asking them about their everyday lives in their segregated schools.

When we reported the latest outrage or a complicated political issue I would telephone the BBC's news chief in Belfast and read him our script to make sure that we, a world away in London, were getting our facts right. It would be so easy to accuse a programme like *Newsround* of twisting the facts to fit an agenda, but to my knowledge no individual or organization ever did that.

During those years, we also covered many significant global events. Among those that spring to mind were the end of the Vietnam War, the massacre in Beijing's Tiananmen Square, the shooting of Pope John Paul II in Rome (another case of 'you heard it first on *Newsround*') and the Chernobyl nuclear disaster. Every one of those stories needed careful thought before we transmitted it.

*

But lighter aspects of life that affected our viewers were also crucial to our agenda, and we reported a long list of 'firsts' in the UK, including the arrivals of McDonald's, CDs ('you can spread jam on them and they'll still play' I said, and proved it), the Sony Walkman and the Pac-Man video game.

We filmed the teenage group Musical Youth going back to their Birmingham school after becoming the youngest people ever to reach No. 1 in the charts. And didn't Paul McDowell and I join up with Madness to record a *Newsround* rap, or is that just a bad dream?

Nick Heathcote, who directed many of my films, certainly remembers the day we drove George Michael and Andrew Ridgeley around London in Nick's gold Cortina convertible. Wham! were at their peak but we had letters from viewers complaining that their fan club was letting them down. It had grown so quickly that it couldn't cope, and young girls who had paid money to join were not getting replies. On *Newsround*, George and Andrew apologized and promised to sort out the problem – which they did, and they were great fun to work with.

Occasionally I almost missed presenting the daily bulletin. Dashing up the stairs to our office from the main newsroom shortly before transmission one day I tripped and spilt the hot soup I was carrying over my face. The duty nurse did what she could and I went on air. On another occasion, I got stuck in the lift. Over the years my accidental companions in the Television Centre lifts had included stars from Eric Morecambe to Olivia Newton-John, but on this occasion there was a receptionist and a lady who, it emerged, suffered from claustrophobia.

Suddenly, the lift juddered to a stop and the lights went out. I had nightmare visions of us plunging towards the basement – after all, that's what happens in disaster movies – but it held steady and I fumbled for the emergency telephone. Almost immediately it was answered by a reassuring voice in the security office.

'Don't worry. Everything is going to be all right,' he said. 'Now, we need to know which floor you are on.'

'I've no idea, but we got on at the fourth' I said.

'But which floor is it now,' he insisted.

'How can I tell?' I replied, getting anxious. 'All the lights are out.'

This pointless conversation continued for a little while longer until, driven by fear and frustration, I slammed down the phone. Immediately I realized my foolishness – you can't cut off your lifeline! So I picked it up again and tried to keep my cool.

The claustrophobic lady was becoming panic-stricken but we reassured her that help would soon be with us. After a while, the rescue team managed to gently lower the lift, restore the power and open the doors. Never have I been so grateful to see a BBC corridor and I made it to the studio with time to spare. That's one story that didn't make the news.

Now and again grown-up television asked me to play with them. One such occasion was the day, in July 1981, of the royal wedding of Prince Charles and Lady Diana Spencer. I had written a paperback for children about Charles and Diana when they became engaged – which, amazingly, an American friend of ours recently found on the internet – so I was well briefed to be part of the BBC One team on the big day.

I began on the night shift, talking to some of the happy thousands camped out along The Mall and getting sensational shots of the sun rising over St Paul's Cathedral. Then I was stationed at Canada Gate, opposite Buckingham Palace, to report live on the comings and goings.

During one of my scene-setting 'spots' I found myself making the kind of stupid remark that makes you want to curl up under a stone as soon as the words are out. As there wasn't much action to report I talked about the lavish display of flowers that had bloomed, so it seemed, almost overnight:

'Just look at these arrays,' I said, 'all dead on cue!'

When the newlyweds were safely back in the Palace our little outside broadcast team was stood down and we relaxed on the grass in Green Park.

The crowd, which was six or seven deep, stayed around to watch the couple depart on their honeymoon. Back at Television Centre, Angela Rippon was presenting a programme that looked back on the morning's events. Suddenly our floor manager got a message. 'The videotape has gone down and we are coming to you very shortly for crowd reactions.'

We scrambled to our feet and I turned to a man with a young family who I'd just been chatting to. 'Can I talk to you on camera?' I asked. 'Sorry, John, you can't – I'm Special Branch. Undercover.' So we ran up to the crowd and as we got there we heard we were on air. 'Didn't see very much,' said one lady, 'as we were right at the back.'

After a couple of other similar uninspired comments I moved to a group who were carrying cardboard periscopes, made specially for the day. Surely they must have witnessed something, but they had such thick Irish accents I couldn't make out what they were saying.

There was a real danger of this developing into one of the worst crowd interviews ever when, thankfully, the video problem was solved and we could relax again. For me, some things were best forgotten on that day to remember.

9. Saturday Mornings

When Danny Cohen took over as the youngest-ever controller of BBC One in 2010 he invited me to his glass-panelled office at New Broadcasting House for a 'getting to know you' chat. Now it's always a little worrying for seasoned broadcasters like me when people who were born long after we started our careers begin taking over the shop. Are they going to play the card which now dares not speak its name – age – and politely show you the door?

After all, Danny had made his name running the youth channel BBC Three and was obviously going far, but would I be part of his future plans? My fears vanished as soon as I opened the door because the first thing he said after shaking hands proved he was a fan from way back – and what he said was a telephone number: 01-811-8055. That, as any child knows if they were growing up between 1976 and 1982 – and that included Danny – was the telephone number of *Multi-Coloured Swap Shop*, the Saturday morning show that helped change the way television interacts with its viewers.

It was a number ingrained in their memory as they tried in their thousands, all morning, week after week, to get through to speak live to one of the guest stars or swap something. I was lucky enough to be one of the presenters. Like Danny and a whole generation of kids watching along with him round the country, I have great affection for the show. Noel Edmonds, at the time the star of Radio One's breakfast show and *Top of the Pops*, was the front man with Liverpudlian teenager Keith Chegwin out on location doing lots of swaps with crowds of children and Maggie Philbin and me helping out in the studio.

We all became great friends (something which does not necessarily happen when strangers are thrust together by television producers), and that camaraderie came across on the show, which lasted for 146 editions over six years.

It was the BBC's first programme, apart from sports coverage and big occasions, to transmit for three hours non-stop and the first children's programme without a script – there was a running order, but we improvised the rest.

Multi-Coloured Swap Shop began after the then controller of BBC One, Brian Cowgill, told Edward Barnes, assistant head of children's TV and the creator of *Newsround*, that he was not happy with the audience figures for Saturday morning. As Edward recalls:

> He said they were all right during the week – in fact ITV were nowhere near us – but they were dreadful on Saturday morning. I told him the reason was we had no money so we put on cartoons that children had seen at least twice already. Cowgill's reply was: 'You come up with the idea and I'll find the money.'
>
> Rosemary Gill, who had worked with me on *Blue Peter*, had had the notion sometime before of a half-hour show based around swapping, because she said children loved to swap things, but it didn't happen because there wasn't the budget. But after Cowgill threw down the challenge I realized we could base a whole Saturday morning round the swapping idea. We'd still have third-run cartoons but we

could also have guest stars and pop videos. And a bit of *Newsround* as well, but done in a lighter way, to add a bit of cement to the show.

Cowgill said yes and I handed it over to Rosemary. The only other contribution I made came from a short-lived show we did a year before, based on personal problems, called *Z-Shed*. It featured telephone calls from children which proved that, unlike many adults, they knew exactly what they wanted to say on the phone. We had lots of auditions for the presenter of *Z-Shed* and who should be among them but Noel Edmonds, who got the job and did it very well. So I said to Rosemary that Noel should do *Swap Shop* and we'd have telephones and that was how it all started.

Rose was the ideal person to preside over the organized, highly professional chaos of Saturday mornings. She was warm and caring with a wicked sense of humour, always calm and very, very well organized. As soon as she heard the show had got the go-ahead she exclaimed: 'We've got the hall on Saturday mornings,' just as a Guide leader might say having persuaded the parish council to let her use the village hall. But with a big wink.

Though we always got on very well, and I had the honour of reading at her funeral service many years later, Rose was initially not happy about me being on the show – she thought there was no place for news on her light-hearted *Swap Shop*. But Edward insisted that I should be the 'grit':

> I told her that if it was all 'candy-floss' the show wasn't going to work. My belief was it would look better if there's something in it that people had to concentrate on.

But then between you and her, and I'll never know how it happened, you crept into other bits of the show and that's one of the things that really made the programme. On *Swap Shop* you were seen as a friendly face but you couldn't be all that friendly on *Newsround*. You couldn't laugh your way through the news. But you could have some fun on Saturday mornings, therefore your popularity increased, and that was to *Newsround*'s benefit.

In contrast to supercool Noel and newsy me, the third member of our on-screen team (Maggie Philbin joined later) was 'cheeky chappie Cheggers' – Keith Chegwin, who sadly died in 2017, but was then nineteen years old and brimming with confidence and bonhomie.

He wrote to Edward asking for a job and was sent to meet Rose. She could see that behind the teenage bravado there was a genuinely loveable character, professional beyond his years. He could have been born to host the outside-broadcast swapping sessions; no matter what the weather (and the show always transmitted during the autumn and winter) you could rely on Keith to emit happiness.

And so it was that on 2 October 1976 *Multi-Coloured Swap Shop* hit the screens – the multi-coloured bit, by the way, came from Rose – she had a different colour for all the ingredients: Noel, Cheggers, me, cartoons, guests, swaps, etc., were all identified by individual hues and she mixed them all up on a board until she had a nice rainbow pattern.

For me, the show was a revelation. I was used to broadcasting on my own from a very small news studio in a Television Centre office block with remotely controlled cameras and just the floor manager and autocue operator for company. Now, in my weekend job, I was becoming part of the BBC's light entertainment output amid all the buzz of showbiz.

I was working in one of the big studios on the ground floor, where everything from *I Claudius* to *The Generation Game* was produced, and when I arrived an hour or so before transmission on the first morning I was overawed by the expectant atmosphere, the noise and the intoxicating level of pure excitement. Something big was about to happen.

Bands were practising, cameras were zooming around, the catchy theme tune was booming over loudspeakers, guests such as the current Dr Who, Tom Baker, and his scarf were arriving, a squad of operators was getting ready to answer the phones and in the midst of it all Noel, cool as a cucumber, was rehearsing the one-minute-long opening 'menu'. And that was the only thing during the next three hours that had been rehearsed.

At 9.30 a.m. the presentation announcer handed over to us and suddenly everything calmed down as the opening titles played on the studio monitors and Noel welcomed millions of children to a new experience. During the first few minutes everything went to plan apart from one very critical factor – there didn't seem to be many calls.

I had my own little news desk away from the main set and the idea was that, after joining Noel at his big desk for the opening of the show, I would go to my position and he would hand over to me from time to time. The thought did cross my mind, *Is this going to be a flop? What if no one calls?*

But there was no need to worry – suddenly the phones were red hot and they stayed that way all morning. Later I was told that halfway through the show the Shepherd's Bush telephone exchange almost imploded because it just couldn't handle the deluge of calls. One was from an elderly lady who complained that children were dialling her home number by mistake. Noel took a call from her live on air, apologized for the inconvenience and invited her on to the next *Swap Shop* to help out on the phones!

After a few minutes, he called out the soon-to-become catchphrase 'Where are you, Keith?' and up popped the young fellow on an outside broadcast camera from a spot close to where one of the BBC's sporting events was being covered that afternoon. To cut down on costs, *Swap Shop* shared the OB unit with the sports department and the first Swaporama was from Cardiff Arms Park. Cheggers would urge viewers living nearby to come along (only, of course, with parental permission) and bring along their swaps. He even told them which buses to catch.

In the studio, the best swaps from child viewers were posted on a big board – a doll's cot for a small music box, a guitar for anything, a soft toy for a cowboy outfit; our youngest consumers were clearing their cupboards and seeking replacements. Guest stars and children who came into the studio to talk about their hobbies brought along swaps as well.

Watching from the sidelines, I spotted a moment when television history was made. A young boy was on the phone to Noel and to win a prize he had to be able to see a picture on his TV screen at home. But the telephone was in the hall and, in the days before mobile phones, the cable was not long enough to reach into the lounge where the TV was.

Noel urged him to push back the door as far as he could

and even stretch out on the floor. The young lad was letting Noel know just how his efforts were going and eventually he managed to catch a glimpse of the screen and in triumph got the answer correct. This (though we didn't call it that at the time) was interactive television being born.

Time flew by and at the end of the show everyone realized we had a hit on our hands and, more to the point, both we and the audience had been having a lot of fun. Nobody cared if a camera got into shot – in fact on a later show a cameraman was handed an L-plate after he accidentally bumped into Noel's desk – or someone got a swap wrong, because it was all part of the relaxed informality.

Afterwards there were glasses of celebratory champagne in the bar, and as I walked home through Shepherd's Bush in a happy mood I spotted a kitten in a pet shop window and bought it. I know you should never buy a pet on impulse, especially as my little girls had been asking for a rabbit, but Lindy looked so appealing that I did my own swap – her in exchange for a couple of pounds.

When I walked in with my surprise box the girls instantly fell in love with the kitten; all thoughts of a rabbit were forgotten and Marilyn took it rather well, considering we had never discussed having a cat. That evening we had been invited by some friends to a drinks party. 'We ought to take Lindy along,' I said, 'otherwise she might feel lonely.' The girls were thrilled, and though Marilyn gave me a look that said *That's the first time I've ever heard a cat being taken to a cocktail party*, Lindy came with us anyway, in the cardboard box from the pet shop.

But, unseen by anyone, she managed to escape. A search was mounted around the house and because the outside door had been left open for a while Marilyn and I called 'Lindy!' up and down the street – not that she would have recognized her name just a few hours after being given it. There was no sign of the little kitten and the girls were distraught as we went home with an empty box. I'd already warned them that, as we lived in a big city, sometimes cats leave home and are never seen again. 'I didn't think Lindy would go so quickly,' cried Emma.

What had started as a really happy day with the birth of *Swap Shop* ended in tears in the Craven household. But next morning our friends phoned to say they had found Lindy. She had crept through a small hole in the back of the settee and was curled up inside, fast asleep. So out came that box again and Lindy returned home – and we kept an extra-watchful eye on her. She was with us for many years as a much-loved member of the family and a lasting reminder of the first *Swap Shop*.

In the following weeks the show consolidated its success and many famous actors, entertainers and pop idols urged their agents to get them on. And Britain's parents could have a bit of a lie-in knowing that their offspring were being safely entertained downstairs. We had quickly become part of the nation's Saturday morning routine. Then, after a couple of months, a wildcat strike (it was a time of great industrial discontent) put *Swap Shop* off the air.

Noel and I and the rest of the team were already in the studio, and with nothing else to do we helped answer calls from confused and alarmed viewers – and not just children. From what we gathered, many parents who could now enjoy a Saturday morning lie-in had been taken shall we say by surprise when their children suddenly burst into the bedroom shouting: '*Swap Shop*'s not on!'

For me, the programme came at the end of a busy week. There was *Newsround* from Monday to Thursday and then on Friday I would regularly make a short film for *Swap Shop*, which we would edit that night and transmit the next morning. Some Saturdays I would get to Television Centre feeling totally shattered, but as soon as I walked into the studio the tiredness vanished – I was getting my *Swap Shop* 'fix'.

Gradually, I became a more integral part of the show. My solitary news desk disappeared and I did all my News Swap spots sitting at the main desk with Noel. As well as big stories of the day I also read out some very local 'news' sent in by viewers – our 'stringers of the week' and took phone calls from them on issues that mattered to them such as homework, friendships and spending money. One letter came from a boy who had lost his teddy bear called Little Ted in a Little Chef on the A5. I appealed to anyone who found it to let me know. They did, and he got his teddy back. Just part of the *Swap Shop* service.

Some weeks I went to the Friday morning production meeting along with Noel and Keith when we would add our two pennyworth to the ideas for the next show, suggest guests and sign autographs. At one meeting Billy Connolly's name was put forward and some around the table though it a bad idea because he was renowned for his colourful language and they feared he might let something slip. Others, including Noel and I, thought he would be great – and we were right. He hit just the right note with his humour, he nicknamed the show 'The Swappie' and was one of our best guests ever.

Afterwards I mentioned to The Big Yin the concerns some had expressed and he was flabbergasted. 'I have kids at home who sit watching *Swap Shop* in their multi-coloured socks and try all morning to get through on the phone,' he said. 'Obviously they have never seen me on stage so this was my chance to show them what their dad does for a living. Do you think I would let them down?'

Dame Edna Everage was another guest, and on her first visit I had a long chat in the make-up room with Barry Humphries before he went off to change. When Dame Edna appeared she had to be introduced to me and be told what I did because we hadn't met – that's how seriously Barry takes the role.

But I was rather flattered to hear that Paul McCartney knew who I was. I was standing in the wings when he arrived in the studio with his young son and one of the floor assistants heard him say: 'Look, James, there's John Craven!' Fame indeed.

When Maggie Philbin joined us after two seasons, the team was complete and she brought a refreshing girl's-eye view and a really warm presence to the show. She and Keith started dating and I announced their engagement on *Newsround*. Marilyn and I were guests at their wedding and they named their daughter Rose, after Rosemary Gill.

My job continued to be bringing some grit to the show but I was also included in the fun. I got dressed as a Bee Gee to sing 'Massachusetts', tested out 'space dust' sherbet on the tip of my tongue with a microphone inches away and had one of my News Swap letters chewed up in front of the cameras by Noel's one-eared goat called (what else?) Vincent. My jumpers had already become famous on *Newsround*, largely because of the novelty of a BBC newsreader not wearing a suit, but on *Swap Shop* they went stellar.

Noel even persuaded me to attempt an unofficial world record for the number of jumpers that could be worn on one person. With his help I managed to get on more than twenty and ended up looking like a very hot Michelin Man. I happened to mention on one show that it would be nice to have my name knitted into a jumper so the director needn't put it on a caption. One over-enthusiastic viewer knitted me a jumper with my name stitched into it dozens of times, covering every inch.

Delia Smith often popped in with an easy recipe for our audience that they could prepare for their family or friends and she was great fun – a real giggler. She even made the teas when a *Swap Shop* team, including Noel, Keith and I, took on her husband Michael's village cricket team in Suffolk and lost.

Pop music was, of course, a vital ingredient on *Swap Shop* and just about every major singer or group came along to talk on the phones to their fans, offer souvenir swaps and either sing live or cue-in the video of their latest release. But I did miss out on the chance to be in the Top 20, for a very good reason.

The idea was put forward that Noel, Keith, Maggie and I should form a pop group with the name – and I can't recall why – of Brown Sauce. But on the day before we were due to go into the recording studio with a song called 'I Wanna Be a Winner' I got a message just before *Newsround* went on air to say that my elder daughter Emma, who was then eight, had been involved in an accident while staying with her grandmother in the north-east.

They had been knocked down by a runaway horse as they walked along a pavement and both had been hurt. The policeman who gave me the message said Emma had a broken ankle and concussion, 'like a biscuit tin being damaged but none of the biscuits broken'.

Marilyn and I hastily caught the evening flight from Heathrow to Newcastle to be with Emma and her granny and fortunately the officer's diagnosis was right. Emma had no serious long-term injury and both she and my mother-in-law were soon on the road to recovery. To cheer her up the surgeon who fixed Emma's ankle also put a matching plaster on her teddy bear – a kindness he may later have regretted. The local newspaper published a photo of Emma and me and the teddy and I later heard that, when word got round, every young patient wanted a plaster on their favourite cuddly toy.

The Brown Sauce recording went on without me and it got to No. 15 in the singles chart and stayed in the Top 40 for nine weeks. Much more important to me, though, was that Emma and her granny survived a very nasty accident. And later I was available for other 'performances' by the four of us. We did a studio panto, *Cinderella*, with Maggie as Cinders, Keith as Prince Charming and Noel and me as the ugly sisters with Rice Krispies stuck to my face as warts.

And we 'acted' our own version of *Star Trek*, called, as you'd expect, 'Swap Trek', in a mock-up of the cockpit of the USS *Enterprise*. I played 'Mr Speck', but with Spock's long ears and flattened hair, and one of my lines was: 'We are heading out of control towards a previously unknown planet and unless checked the total destruction of our team is inevitable and *Grandstand* will just have to start earlier.' Always the cheerful one, even in outer space, and I must admit I was a little concerned that, with the help of the make-up department, I bore more than a passing resemblance to Leonard Nimoy.

Even more worrying was the time Noel and I got lost on the way to the studio. By then I was living not far from him in rural Buckinghamshire, and on this particular Saturday he gave me a lift. There were problems on the A40 and I thought I knew a way through the back doubles, but I was wrong. Don't forget these were the days before satellite navigation. We seemed to be going round in circles until finally I spotted a road I recognized. We reached Television Centre with minutes to spare.

The list of stars who queued up to be on the show was like a broadcasting Hall of Fame. Here are just a few who spring to mind: David Attenborough, Cliff Richard, Joanna Lumley, Paul Daniels, Debbie Harry, Christopher 'Superman' Reeve, Bonnie Tyler (her song 'It's a Heartache' was interpreted by Noel as 'It's a Hard Egg'), Michael Crawford, Penelope Keith, Rod Hull and Emu, Elvis Costello, David Bellamy . . . and there were hundreds more.

Another 'first' for the show was when Noel left the studio to go on walkabouts around Television Centre to see what was going on elsewhere. It caused quite a rumpus among the old-stagers who couldn't believe the hand-held cameras were following him into unlit areas and wandering at will among the sets of some of the best-loved sitcoms and dramas, meeting anyone who happened to be there – including major stars busy rehearsing.

Noel went into the control gallery of *Swap Shop*, from where the show was directed, and the make-up room so viewers could see what was going on behind the scenes, and on one occasion I took over from him and did a guided tour of the main newsroom.

Once again, *Swap Shop* was setting the pace for live, as-it's-happening broadcasting. But when I look at surviving tapes of the show (which, sadly, are few in number) I can't help but notice how slow the pace was compared to current entertainment productions, even though back then we thought it was pretty hectic. Children were given lots of time to express themselves on the phone and the guests had perhaps fifteen minutes to chat, take calls and offer swaps.

These days they'd be lucky to get five minutes max before there's the relentless need to move on to something else. As always, I blame it all on the remote control. Back in the day when all viewers, not just children, had to get out of their chairs and walk across to the set to change channels there was no incentive to flick from station to station. My theory is that the 'zapper' is responsible for the attention span deficit that so dominates the policies of modern programme-makers.

No sequence can last more than about thirty seconds otherwise viewers might get bored and hop to another channel. That's their thinking, and I long for a return to the practice of things being allowed to run their natural course – but I'm afraid it will never happen.

Celebrities weren't the only Saturday morning stars. We also had all the characters from the cartoons and the soft toys that littered the main desk, most famously Posh Paws the purple dinosaur, whose name was Swap Shop spelt backwards (almost).

And I 'had a hand' in another one, Lamb, and that all started because I was sitting next to Noel when the machine about to transmit the next video broke down. We needed to fill in so, unseen by the viewers, I grabbed one of the soft toys, which happened to be a lamb, and ducked underneath the desk. Then I slowly raised its head and upper body into shot on the desk and Noel had a long, one-sided conversation with it until he got the signal that the video machine was working again.

As a result Lamb became a bit of a star, making regular appearances. It was modified so that I could put my hand up into its head and front legs and make it move a little, nod its head and wrinkle its nose. We even made a film about how Lamb was reacting to its unexpected fame, with Noel's mum agreeing to put on a chauffer's hat and drive our new star around the countryside in a Rolls-Royce. As they passed our local primary school, my seven-year-old daughter Victoria was among the excited pupils waving and cheering.

On the final *Swap Shop*, in March 1982, I crawled from under the desk with Lamb on my arm and my until-then secret role was revealed. *Swap Shop* allowed me to be both serious journalist and silly puppeteer and I think the millions in our young audience rather liked the combination. Lamb may have had a passing moment of fame but it is not forgotten by the Craven family. After that farewell performance it came home with me and every Christmas still plays a starring role, at the foot of our tree.

Of course, there was another children's show on Saturday mornings. We had an anarchic rival in *Tiswas* on ITV, with Chris Tarrant, Sally James, Lenny Henry and the Phantom Flan Flinger. Each programme had its own loyal fans.

When *Swap Shop* finally closed its doors, I thought it would be the last time I would ever see the set which had become so much part of our lives. But thirty years after the very first show was aired, BBC Two made a programme looking at the history of Saturday morning television called *It Started with Swap Shop*. Noel, Maggie, Keith and I walked into the studio and there in front of us, precisely recreated, was the original yellow and orange set. The theme tune was playing, the big clock was set at 9.30, emblazoned across the desk was 01-811-8055 – and there was a big lump in all our throats.

I had stayed friends with the others, even though Noel 'did the dirty' on me when he moved to Saturday evenings with his hugely popular *House Party* show and handed me one of his Gotcha Oscars. This was the section of the show when he made well-known people look rather foolish. And he really tricked me – with the help of Marilyn and his then wife, Helen.

From time to time we went out as a foursome for a meal and on this occassion Noel suggested we try a restaurant in Princes Risborough that had just been opened by a mutual friend. For me it went badly from the start; my chair collapsed (electronically controlled, I later discovered), my wine glass leaked (a hole had been drilled in it), and my meal was more or less inedible – the pasta dish was incredibly spicy, my roast duck was uncooked and my salad was served to Noel, who discovered it had a slug in it.

But as I was feeling charitable I put it down to early glitches in our friend's new venture – and anyway the wine was fine and flowing freely, a lot of it down my arm, and I wasn't driving but was enjoying a good old natter.

Then to my astonishment I spotted Tony, one of my happily married neighbours, dining up-close and personal with a very attractive young lady who was not his wife. When she went to powder her nose I couldn't resist going over and asking him who she was. 'An old friend,' Tony said, rather sheepishly. 'She doesn't look that old to me,' said Noel when I reported back to our table.

I was still getting over that surprise when smoke suddenly began billowing through the kitchen doorway close to where we were sitting and in no time at all the fire brigade arrived. Amid all the confusion Noel handed me a phone with a recorded message from him saying: 'Gotcha'. The whole thing, including my neighbour's 'friend', the other diners and the firemen, had all been part of the con and half the dining area had been screened off to hide the cameras.

Looking back on it now, I can't believe I had been so gullible. Maybe it was because I never believed my old pal Noel would 'get' me and also because our wives – who both normally shied away from television cameras – were happily playing along with this very elaborate charade.

Noel stitched up well over a hundred celebrities with his Gotchas, and when it came to the last-ever *House Party* he'd heard rumours that, quite rightly, some of his victims were planning to get their own back. He suspected it would be at the West London hotel where he always stayed before the show. So the day before, he phoned Marilyn at home to ask if he could stay. I'd long ago forgiven him (and Marilyn and Helen) for the Gotcha! but I bet he spent a very jumpy night in our spare bedroom, wondering if somehow we were in on a conspiracy.

As it happened his ploy worked well and he escaped largely unscathed (apart from Freddie Starr spraying him with a fire extinguisher!) on that final *House Party*. At the end Noel did a little sketch with Keith and me in which it appeared we were all back together on *Swap Shop* and the *House Party* had just been a dream – albeit one which, like *Swap Shop*, had a huge impact on the way light entertainment programmes were made.

Back in 1982 the BBC had to find another show to fill the chasm left by *Swap Shop* and, following the age-old adage of 'don't fix it if it's not broken', came up with something very similar: *Saturday SuperStore*. With Noel moving across to prime-time television the producers chose another Radio One breakfast DJ to fill the gap, the laid-back Mike Read, who quickly settled into the job of 'store manager'.

Assisting him was Sarah Greene, lately of *Blue Peter*, and ex-footballer David Icke, who later became notorious as a soothsayer and conspiracy theorist who believed reptilian humanoids wanted to take over the world. In those days, he was just in charge of the sports department. Also on hand were a trio of *Swap Shop* survivors: Maggie, for the first series, Keith as the delivery boy, and me in customer services.

Mike always had his guitar close to hand, and when he wasn't DJing and doing the telly he was in a rather good band. In 1981 the Craven family had moved thirty miles from West London to the Chiltern Hills in Buckinghamshire. Mike and his band agreed to play at a couple of charity events in our village and went down a storm – and Mike didn't charge.

So Saturday mornings carried on much the same as before, and for me the absolute highlight came shortly before the general election of 1987 when all three political leaders – David Steele, Neil Kinnock and Margaret Thatcher – accepted our invitation to appear on separate weeks. They agreed to answer questions from viewers on the phone, and take part in some of the fun elements, like giving their verdicts on the latest pop singles.

It was Mrs Thatcher who, of course, made the big headlines because it was the first time a serving prime minister had appeared live on such a show. But this was not her debut on children's television. Fourteen years earlier she had been a guest of Valerie Singleton on the series *Val Meets the VIPs*.

At that time she was Minister of Education and had been given the nickname that was to haunt her – 'milk-snatcher' – because she ended free school milk for the under-sevens. In answer to one question from Val's young panel she memorably replied: 'I don't think there will be a woman prime minister in my lifetime. I would not wish to be prime minister.'

Bearing that in mind, and with politicians and the media having long memories when it comes to things that maybe should not have been said, I wondered how circumspect she would be on *SuperStore*. She arrived at Television Centre wearing a fawn suit and pearls, hair immaculate as always and carrying a wicker basket in which she had gifts to offer to the audience at home.

I went to greet her in the main reception area and as her high heels clickety-clacked across the hard floor she reminded me of a very well turned-out granny arriving at a children's party.

On the show she chatted with Keith and Sarah and when she joined the pop panel with Mike probably damaged the career of one emerging group by saying she liked their music and giving them four out of five. But she was merciless about the Style Council's latest. 'It didn't go anywhere and I wanted to say "get on with it". It was too much like a rehearsal.'

Then came the moment when she faced young inquisitors from across the nation, a bunch of mini Robin Days with a direct line to the prime minster who were standing by with a barrage of questions ranging from nuclear war to family life to the secrets of the Number 10 door. I sat with her in front of a group of youngsters in the *SuperStore* coffee shop and put through the calls. Like her, I had no idea of what was coming next.

Ben Cronin, aged twelve, asked her what was the most important decision she had ever made. 'I think the most important decision was to stand up to people who threaten you, whether with terrorism, strikes or any form of intimidation,' she said, adding that children should stand up to playground bullies.

When Suzanne Thurston, eleven, wanted to know if she discussed political problems with her family, Mrs Thatcher told her they kept her in touch with public opinion. 'My husband is very frank,' she said, 'and when you're down in the dumps families are marvellous – they say "Come on, Mum, cheer up." It's often what I say to my cabinet at the end of a particularly difficult meeting: "Come on, cheer up. Life goes on."'

Some of the questions 'from the mouths of babes' were ones that adult professional interviewers might not have thought or dared to ask.

When told she seemed to shout a lot in the Commons she replied: 'I have to because sometimes the people who don't like my answers make a noise. It's much noisier than it used to be.' Her secret for staying calm when under attack was to say to herself: 'Keep cool.'

Stewart Jones, twelve, from Doncaster, urged her to reveal the date of the next general election but she wouldn't be drawn on that one. Kellie Bunting wanted to know if it was true that the door of Number 10 could only be opened from the inside. Mrs Thatcher confirmed it for her, saying they couldn't let just anyone in and callers were vetted by closed-circuit TV before crossing the threshold.

She seemed to be enjoying this junior version of Prime Minister's Question Time and watching her in action at close quarters I realized what an accomplished television performer she was. At one point, she told me about the special gift she was offering as a prize – a small piece of English porcelain similar to ones she gave foreign leaders when they visited her at Number 10.

And as she pointed it out, with five cameras looking at us, I saw her glance up to see which one had its red light on so she could hold it to that camera for a close-up. *What a professional*, I thought.

The questions kept flowing, from unemployment to education, but it was left to fourteen-year-old Alison Standfast from South London to, quite literally, drop the bombshell. In a clear and determined voice she asked: 'In the event of a nuclear war where will you be?' There were gasps around the studio, including one from Mrs T. 'My goodness me,' she said. 'Now look – the whole point of having nuclear weapons is to stop a war of any kind.'

But Alison, Standfast by nature as well as name, would not be fobbed off with that standard reply. 'But if there is one where will you be?' she demanded. 'I will be in London,' the prime minister told her. To which Alison, undaunted, replied: 'Have you got your own bunker or something?' In true political style, Mrs T dodged that one, saying: 'Let me again point out that the possession of these weapons has kept the peace for forty years.'

And so we moved on to the next caller and it was left to the Sunday newspapers, which gave full coverage to this unique exchange, to reveal that if there was a nuclear attack the prime minister, the entire Cabinet and top civil servants would be ensconced in a warren of bombproof bunkers under Whitehall and lead the country from a war-room equipped with living quarters and a hospital.

When the *Mail on Sunday* gave that information to Alison at her home her reply was: 'I don't know why Mrs Thatcher couldn't have told me that herself.'

At the end of her ninety-minute appearance on the programme, the prime minister said: 'I have loved it, every minute.' But shortly afterwards in the green room I discovered that her husband, Sir Denis, was far from pleased with Alison and her questions.

Holding his trademark gin and tonic, he accused me of 'setting up' Mrs T. 'That question was deliberately planted,' he fumed. 'You could tell that because all the other children were respectful. My wife was tricked.' I told him that was most certainly not the case. All the questions were absolutely genuine and selected from hundreds that were phoned in to provide a fair balance.

And this was proved beyond doubt nearly twenty years later during that special programme to celebrate Saturday morning television. The researchers managed to track down Alison, now a mum and unsurprisingly using her cross-examination skills in the legal profession. She was in the audience and I told her about Sir Denis's allegation that we had colluded over those questions to Mrs Thatcher. 'Not at all – it was all my own work,' she affirmed.

So why did she want to know about the bunker? 'It seemed very important at the time. There had been a film called *When the Wind Blows* [based on a Raymond Briggs story about a nuclear war] and it was a matter close to my heart.' So she had picked up the phone and a few minutes later, to her surprise, was challenging the prime minister and making headlines.

It was a perfect example of those unpredictable, interactive, groundbreaking and hugely enjoyable Saturday mornings that were part of my life for ten years. But when *SuperStore* came to an end, it was time for me to go as well. I had by that time become editor of *Newsround*, and with all the extra work involved there simply was not the time to have fun on the telly at the weekend.

10. *Newsround Extra* Bangladesh

Within weeks of seeing my newborn daughter Victoria come into this world, with all the care and medical professionalism that we take for granted in the UK, I found myself in a country where babies were dying in front of my eyes from starvation and disease. The contrast was unbearable. I knew Bangladesh was beset by many problems – poverty, cyclones, floods and famine – because, after all, I had reported them on *Newsround*. But somehow I didn't expect it to be this bad.

I had to take a grip on my emotions and, together with director Mike Beynon, work out how we could report to our audience back home on the realities of life for children in what was then the country with the world's youngest population – more than half of all Bangladeshis were under the age of sixteen. Obviously we would not show babies on the brink of death but we could allude to the high mortality rate and film the positive actions being taken by some of the eighty-odd charity organizations based in Bangladesh to save as many children as they could.

We were there because, with *Newsround* now firmly part of the schedules, our team had put forward the idea of a spin-off, a current affairs/documentary strand that would give us the chance to explore in greater depth many of the subjects we could give only a couple of minutes to on the bulletin. The bosses agreed and *Newsround Extra*, with some editions running at thirty minutes and others at ten, first hit the screens in the summer of 1975. Two of the opening half-hour themes could not have been more different – Bangladesh and the Bay City Rollers.

The Rollers were to the mid-seventies a slightly less frenzied version of what the Beatles were to the mid-sixties. They were five bouncing young Scotsmen who wore tartan and sent their teeny-bopper fans wild. The press called it Rollermania because everywhere they went they were mobbed by hysterical girls who, like their heroes, clad themselves in tartan trews, hats and scarves – the first time that fans had been so uniquely identifiable.

'We all like the tartan gimmick,' fourteen-year-old Lucy told me, her scarf tied round her wrist. 'I love the Osmonds as well but you can't go round wearing the American flag, can you? You'd look stupid.' Like all the others, she'd plastered her bedroom walls with photos of the group, and though she had tried to get to concerts she had only seen and heard them on television.

The boys – Alan and Derek Longmuir, Stuart 'Woody' Wood, Eric Faulkner and Les McKeown – had just had a No. 1 in the charts with 'Bye Bye Baby' when I met them in their dressing room at a Bristol theatre at the end of a hugely successful UK tour.

They were chatty and likeable in a boy-next-door way and when I asked Derek to explain their impact on the fans he told me: 'We're just ordinary guys, really. We dress cheaply so they can copy us and we are about the same age as them so we can communicate.'

Outside, thousands of girls were screaming for them and most didn't have tickets. They were there in the hope of catching a glimpse of the Rollers as they leapt from a nondescript white van into the theatre. It took them two seconds. 'I saw Woody's leg,' said one of them. 'Musically, they don't impress me,' I was told by Brian Harrigan of *Melody Maker*, 'but are we talking about music or the Bay City Rollers? Does music really matter when 3,000 girls are screaming and you can't hear what they're singing?' Just like it was in Bradford, I remembered, when the Beatles hit town.

The Rollers were a music phenomenon and there must be many middle-aged ladies today who remember with fondness the days they donned the tartan and dreamt that their boyfriend might be Woody or Les or Stuart . . . It was all harmless and most of the girls who collapsed at their concerts were cured with a whiff of smelling salts. Rollermania was short lived, but it was only right that we covered it when it was at its height – and it was only right, too, that we told our audience about children who would never have the opportunity to enjoy that kind of rather frivolous experience: children who lived in places like Bangladesh.

Mike and I flew to Dhaka, the country's capital, the long way round – via Moscow on Aeroflot because it was the cheapest ticket. Our budget was limited, unlike the charity we were working with who sent their team direct by British Airways. When we walked out of the airport building Dhaka hit us for six. The heat and humidity, the vast surge of people all seemingly going somewhere, the noise, the smell, the poverty, the bicycle rickshaws – it was our first visit to the developing world and it took my breath away.

Because we wanted to see Bangladesh at first hand and be on the scene if anything happened we opted, with the agreement of our two-man film crew from BBC News, to stay in a small bungalow near to where the aid agencies were operating instead of in a big, air-conditioned hotel. It meant we sweated for twenty-four hours a day and ate a lot of local food but to my amazement I didn't have a single tummy upset. Since then I've been to many developing countries and have two golden rules: One – eat what they eat. Two – whisky kills all known germs.

One of our first jobs was to calm down the crew. Both were news veterans who'd spent years on foreign assignments where every night meant going to the local airport and trying to get that day's film cans shipped back to London on the first available flight (and that often meant trying to persuade a passenger to hand-carry them) because there were no satellites.

For the first time in their careers they didn't have a deadline to meet – we were filming for nearly three weeks and would be taking the film cans back with us at the end of the trip, so they could relax a little in the evenings.

Everywhere we went there were children and they were fascinated by us and our equipment, especially when the sound recordist changed rolls of film in a big black bag. They gathered round expecting some kind of magic trick and he put on quite a show for them.

The country was still recovering from a bloody revolution only three years into its life. The ruler, Sheik Mujibur Rahman, and almost his entire family, including his wife and three sons, had been assassinated in a *coup d'état* and there had been a short-lived counter revolution. So we arrived in a jumpy, insecure country, but no one threatened or impeded us as we travelled around – in fact, we were welcomed with smiles everywhere. Most of the humanitarian work we came across was carried out by international aid agencies, which left the government, theoretically at least, with the opportunity to build an efficient infrastructure.

So at first sight it looked as though Bangladesh was being run by the United Nations, Save the Children, et al. – their vehicles were everywhere. And we came across many hopeful things to report: clinics were reducing the death rates among babies and more parents were being persuaded to send their children to free primary schools, a hard decision for them to make because until then the children had worked all day to boost the family income. Destitute women were being given the chance to help their families by toiling in the fields, and keeping much of the produce.

We went up-country on the romantically named Bengal Night Express, which was far from romantic and travelled very slowly, perhaps fortunately, for most of the passengers were clinging to the roof and doors. The conductor put a thin blanket on my wooden bench, which made it first class, and bolted the doors and windows to stop attacks from dacoits (armed bandits) along the line. As we left Dhaka station countless children, some without clothes, were sleeping on the platform – yet another of many, many images during that trip which forcibly reminded me of just how fortunate my own children were. On another journey we sailed up the wide Jamuna river to film Save the Children projects along the banks.

As we passed, people gathered and waved and, seeing our camera, called out '*Zindabad*', which means 'long live' and sometimes even 'Mark Tully *zindabad*', which proved how much the BBC's famous correspondent in the subcontinent was respected, even in the poorest of places.

One night we stayed with a retired tea planter who had

lived there all his life and as we sat down to dinner it was as if the days of the Raj had never faded away.

The fine tableware and the servants were a glaring, uncomfortable contrast to the village huts we had filmed, where people had few if any possessions and ate with their fingers. Our host warned us that at nine o'clock all conversation must cease. That was when he switched on his ancient radio, already tuned to the BBC World Service in London.

Listening to the news was his nightly ritual, with the overhead fans whirring and the cicadas chirping outside. We waited with him in anticipation – we hadn't heard any news ourselves for days – but it was one of those nights when nothing much seemed to be happening in the world. The lead item dealt with the third day of a minor uprising in South America and not even the great Mark Tully had a story for us.

Some years later, I was in the headquarters of the World Service at Bush House in London and was allowed, as a special favour, to say those three famous words 'This is London' to the microphone during the couple of seconds before the signature tune 'Lilliburlero' and the Greenwich time signal heralded the hour.

Vast millions were listening around the globe and as the moment approached I felt the nerves coming on. There were so many permutations of that small but mighty sentence. THIS is London; This IS London; This is LONDON; THIS IS LONDON. Then I remembered the old tea planter sitting by his wireless with the signal ebbing and flowing and opted, in a stentorian way, for 'This is LONDON'. I'm sure he would have approved.

As we criss-crossed the verdant, beautiful and largely flat

landscape of Bangladesh we focussed almost entirely on its children, from a little boy who collected pieces of scrap paper for a living to a thirteen-year-old mother who seemed far older than her years and no wonder, with such responsibility thrust upon one so young.

Others told me how they had lost their families or their homes or both in the recent floods. What shocked me was how matter-of-fact they were; there was a kind of fatalism in their lives but, as Mike noted, they didn't seem to carry their sadness with them. They still played their games and sang their songs, perhaps because they had no way of comparing their lives with any others, especially in the privileged West. Goodness knows what they would have made of the Bay City Rollers and their screaming, tartan-wearing fans. They lived for today and probably did not think about tomorrow. We left the country feeling that, given half a chance – if natural disasters and political turmoil could be avoided – things just might get better.

On our last evening there came a culture shock even greater than the one we had experienced with the tea planter. We were invited to a smart soirée in Dhaka hosted by government officials and amid all the polite chit-chat on a veranda one lady in an elegant sari asked me: 'Have you seen any of our poor people yet?' 'Nothing but, ever since we got here,' I told her, and she looked rather shocked.

The state-run national newspaper covered our visit, putting a few choice words in our mouths about how we had found no sign of lackadaisical behaviour (what was that all about?) and I, John Krelin, was quoted as saying that Bangladesh 'was now standing on its own foot'. A charming error by the printers, but I did leave the country hoping that the world's youngest nation was taking its first hesitant steps away from its dependence on international charity.

Despite all the deprivation and uncertainty that Bangladesh faced we enjoyed our stay in this emergent land and produced two films which I think gave an honest account to our audience of what life was like for people their age who had none of the advantages they had. A few months later, I'm proud to say, BAFTA gave us the Harlequin Award for best children's factual programme of the year.

Eight years later I was back in Bangladesh for another *Newsround Extra* and, though not much seemed to have changed, I did witness what was the nearest thing to a miracle. During her state visit there in 1983 the Queen went to the Save the Children intensive-care clinic in Dhaka and was visibly moved by the plight of two little children when she looked into their cots. Both were mere skin and bones, hardly clinging to life, and film of the poignant encounter was shown on news bulletins around the world, including *Newsround.*

But what happened afterwards? I was determined to find out. These two destitute infants, whose names I discovered were Fulbanu and Razzak, had been thrust into the spotlight because the Queen was with them for a few brief moments. By following up their story, for good or bad, we could show our viewers the work of this remarkable clinic. Every year more than 5,000 mothers found their way there, often walking for many miles, after hearing by word of mouth that it would make their sick babies better. Its workload had more than doubled since my first trip to Bangladesh.

When I reached Dhaka six months after the royal visit I learned that Fulbanu, now three, had pulled through and so had two-year-old Razzak, unlike some other little ones in the clinic on the day the Queen called in. Three weeks of special feeding had put flesh on their bodies and a sparkle into their eyes, and while they were being treated their mothers, like all the others who seek help, had been shown how to make low-cost nutritional food and a fluid which would help fight diarrhoea – a major killer of children.

Fulbanu and Razzak had been allowed to go home and, thanks to the Save the Children team, I managed to track them down. Pretty little Fulbanu was unrecognizable from the footage of her in the cot; she had blossomed and was a bundle of health. And Razzak was even more of a bundle – he was positively chubby.

I took them both back to the clinic for check-ups, which they passed with no problems and miraculously they were quite a weight when I held them in my arms. Fulbanu and Razzak were just two of many thousands of children that aid organizations were saving from death and six years later *Newsround* showed our 'continuing concern' over issues that we highlight, and try not to forget, by returning to Bangladesh.

Our reporter Roger Finn travelled to Razzak's village and found that he was the picture of good health, if a little confused by the attention he was receiving from a foreign film crew. 'The whole village turned out to greet us and Razzak just couldn't understand what all the fuss was about,' says Roger.

But to *Newsround*, Razzak was a very special boy. Sadly, the aid workers had lost contact with Fulbanu. In their innocence, through that chance encounter with the Queen and later with *Newsround*, the two of them had shown the children of Britain what can be done to help those less fortunate in distant lands.

11. Operation Noah

The Pan-American Highway is the world's longest road. It runs non-stop for 30,000 kilometres from Alaska to the tip of South America apart from a one-hundred mile break in Panama. There is a very good reason for this gigantic 'hole in the road' – a dense swathe of untamed, impenetrable tropical jungle known as the Darien Gap. And that's where I found myself.

Before leaving London I'd had all the necessary jabs but nothing prepared me for the experience of living rough in a rainforest: the stifling heat, the tree-top canopy blotting out the daylight, the ever-present danger from creatures great and small, some of which I'd never heard of before. It was one of the last remaining truly wild places on earth and I was there on a mission.

Mike Beynon, who later had a distinguished career producing natural history programmes, including *The Really Wild Show*, had read about a project being planned by the International Society for the Protection of Animals (ISPA) to rescue thousands of animals in the Darien Gap from the rapidly rising waters of a new man-made lake. It was to be called, appropriately, 'Operation Noah' and in the safety of the *Newsround* office we thought it would be rather a good idea if we joined in and made a film for *Newsround Extra*.

Our offer was accepted by ISPA's field director, John Walsh, and a few weeks later we drove down the Pan-American Highway until it came to an abrupt stop. In front of us, all the way to Colombia, the next country south, was a thick blanket of rainforest. John was there to meet us and guided us through the jungle to his island camp, which had, until the waters began engulfing it, been the top of a hill.

The cause of this wildlife crisis was a new hydroelectric power station being built to keep the lights twinkling in distant Panama City. A huge dam had been constructed and its turbines needed a massive force of water to produce electricity – so the River Bayano and its thirty tributaries were blocked and the new lake, sixty miles long, was forming.

As the level rose, jungle trees were drowned and so were animals that could not escape. 'The indigenous Cuna people were warned and moved to safety above what would be the final water level but no one could warn the animals,' John told me. 'That's why we are here. Our target is to save them and it doesn't matter if they are dangerous or whether they are as common as a sloth or as rare as a naked-tailed armadillo. An animal doesn't know whether it's rare or not. It just senses it's in danger and our job is to take it out of danger.'

John, a big, burly and hugely dedicated American botanist, had recruited a team of Cunas and fitted outboard motors to their pink-painted dugout canoes. Every day they set out onto the lake searching for creatures which had climbed in desperation to the topmost branches of trees or were struggling in the water. John allocated me a canoe and took me out on the water straight away.

We'd been skimming along for only a few minutes when I spotted a coatimundi – a small, racoon-like animal – in a desperate state. But how could it possibly understand that

I had come from afar to be its saviour? So it did its very best to thwart my rescue attempts as I tried to scoop it up with a large net. Eventually, after much fumbling and advice from John in his canoe, I did manage to trap it. Later, like all the other 'victims', it would be released into a safe habitat. It was my first of many contributions to Operation Noah.

I quickly learned from the Cunas that the best technique was to hurl my canoe into a semi-submerged tree, shake the branches as vigorously as I could and then grab whatever creature fell out – within limits. The big fear was that a deadly snake might join me in the dugout and there were lots of them in the Bayano. Most feared by the Cunas was the feu-de-lance – large, aggressive and quick to strike, it was responsible for half the snake bites in Central America. On one of my tree-shaking missions I spotted what to me looked suspiciously like a feu-de-lance on one of the branches. I took no chances and swiftly moved on.

Even life in the base camp could have its risky moments. One evening we were sitting round the campfire when I felt something crawling up my bare leg. 'Don't move, it's a tarantula!' yelled John and dashed off to his tent. *What's he doing*, I thought, *deserting me in this moment of need!* But he quickly returned with a stout glove and deftly plucked the large, hairy spider from my leg. Tarantulas don't kill humans but they can give a nasty bite and that was one creature I didn't save. It was dipped into formaldehyde and I have it to this day as a souvenir.

Alongside it in my office at home is my other memento – a long, carved stick rather like a walking stick, coated with a red dye and presented to me by the Cunas. I felt really touched because it entitled me to vote when they held their councils – if only I could understand what they were saying. A few years ago Marilyn and I were on a cruise that called into the San Blas islands off the Panama Coast, which turned out to be another Cuna settlement. As we wandered round a village, a group of men were holding a rather animated meeting and I felt like saying I'd join them if only I had my voting stick with me.

Operation Noah saved a menagerie of species from drowning: sloths, anteaters, monkeys, small deer, lots of snakes, tapirs and even jaguars joined the list, but only one of them became a pet – a white-lipped peccary or wild pig called Gregory Peccary (what else!). In case you are too young to remember, Gregory Peck was a Hollywood film star who made many great movies, including *To Kill A Mockingbird*. Our young peccary became the team's mascot and even joined in some of the expeditions because he didn't like being left alone in the camp. Gregory Peccary obviously loved humans because he would suddenly rush up, his spiky little body almost knocking you over in a grunting gush of friendship.

The camp consisted of a few huts, a kitchen and the pens where rescued animals could be kept until they were released on higher ground (setting them free was, by the way, almost as hazardous as catching them).

We slept in hammocks under the stars, leaving our boots on the ground, and remembering to make sure snakes or nasty insects hadn't slipped into them during the night. In the middle of the camp were three small totem poles put there by the Cunas and under these poles they had placed some of the propeller blades they had smashed when they raced their canoes through floating debris on the lake. Maybe they were hoping their gods would repair them.

So late one night John and I removed the broken blades and replaced them with new ones. Next morning there was great excitement as the Cunas, thinking some kind of miracle had happened, dashed off to fix their outboards. That day they were even more foolhardy on the lake and many more damaged blades ended up under the totem poles. Realizing our mistake, the 'gods' took no further action and from then on reckless canoeists had to explain their behaviour to John.

He had seen a remarkable change in the Cunas – they were still hunters, but during Operation Noah they were using their skills to save not to kill. John told me: 'When I first came here I met the leaders and explained I needed their help but they didn't really understand why until the first rains came after the dam had been closed. They saw animals clinging to floating logs and later . . . when they began to see dead ones in the water, they came to me saying: "We have to go faster."'

Every day we spent eight to ten hours on the lake, returning at dusk with canoe-loads of bedraggled and starving animals. We were wet, tired and hungry but the animals were always tended to first. John's rule was: 'When the animals have eaten, we eat.' Some of those captured were taken to a research station further down the lake where they were checked, measured and tagged before being released into a reserve where scientists would keep watch on them to check how they had survived the ordeal.

I survived with a few bites and scratches and as I headed back up the Pan-American Highway it felt as though I was time-travelling thousands of years from the rainforest to the modern world. But then I realized it was the needs of the modern world that had created this bungle in the jungle and should it happen again in some remote wilderness, in all probability there would be neither the time nor the funding to set up another rescue on the scale of Operation Noah.

12. Mother Teresa and Fatah Singh

It's not often that someone who opens the door to you later becomes a saint. But that happened in the backstreets of Kolkata and the hunched little lady who answered my knock was Mother Teresa. Finding my way to her house in the crowded, suffocating slums was not the only problem I had on that trip. The Communist-led state government of West Bengal seemed reluctant to give *Newsround* permission to film this world-famous nun. The reason, I suspect, was because it resented the constant attention that her presence in the city focussed on the plight of Kolkata's poorest of the poor.

Director Nick Heathcote and I arrived in the city in the mid-1980s after filming healthcare projects elsewhere in India and investigating the threat of pollution damage to the Taj Mahal. The beauty of the Taj takes your breath away but grit and smoke from nearby factories was starting to bite into the glorious white marble of the edifice that Shah Jahan built nearly 400 years ago to show his love for his wife Mumtaz, who had died in childbirth.

As we approached its famous outline in the city of Agra our driver, Kartar Singh Kelly, said wistfully: 'This is the teardrop on the face of India.' I don't know if he made that up himself but it was exactly the sentiment we needed, so we quickly stopped the cab and got him to record the words for us. His feelings were echoed a few years later when Princess

Diana posed for the press sad and alone in front of the Taj – an early indication perhaps that her marriage was on the rocks.

When I asked him about his name he smiled and said: 'I'm one of the Delhi Kellys. My ancestor was a roaming Irishman who settled here, for a while at least, and married a Sikh.' One thing was for sure, Kartar had certainly inherited the blarney.

After the opulence of the Taj Mahal came the devastating contrast between wealth and poverty in Kolkata, with shanty towns lining the route from the airport to the city and fine colonial buildings at its centre. One of them was the Writer's Building, named after the clerks of the British East India Company for whom it was built and now, as we discovered, home of the government secretariat.

It was where we waited and waited (discovering at first hand that one of Britain's legacies to India is bureaucracy) for two days in surroundings which were neo-Dickensian with clerks sitting at high desks in dusty rooms, and could that be a quill I saw?

Maybe not, but the frustration of waiting hour upon hour for a filming permit (one we thought had already been granted) began to play tricks on the mind. Eventually our patience – and the promise of a bottle of whisky to one of the clerks – was rewarded and off we set in search of Mother Teresa at the humble headquarters of the religious order she had founded, the Missionaries of Charity.

We had contacted her office by phone before leaving London and been told we might not be able to film her because she was exhausted after a visit to New York, where she had addressed the United Nations. However, she would try her best to see us 'because it is for the children'. So we took a chance – and she kept her word.

We had to be at the Mother House just after daybreak, and to my great surprise she was there to answer the door. My first impression was of how small, frail and gaunt she was, and when she spoke it was in the faintest whisper – quite a problem for our sound recordist. We walked through to an indoor courtyard where, lit by early shafts of sunlight, a small group of nuns were washing their spare distinctive saris – white with blue edging – in tin buckets. 'These are our only possessions,' Mother Teresa told me. 'Two saris and a bucket to wash them in.' In sharp contrast to the bustle in the streets outside the courtyard was silent and serene and timeless.

We did an interview – it had to be brief because of her state of health – in which she said she had no problems dealing with poverty because she could feed people who were hungry and give them comfort and shelter. 'What concerns me is the poverty of mind and spirit in the countries of the West,' she said. Though she didn't really develop on that, it was a reflection on her opposition to abortion and contraception and what she saw as the breakdown of family life in the developed world. 'That is much more difficult to deal with.' She ended our conversation by giving a blessing to the children of Britain.

Though it had lasted only a few minutes, it was perhaps the most extraordinary interview I have ever conducted because of the aura that surrounded this remarkable woman. She ran a worldwide enterprise from this nondescript house – an enterprise whose only commodity was mercy – and Nick and the crew and I felt spellbound. We agreed later it was the nearest we would ever get to meeting a saint and, as it turned out, we were right. We had been in the presence of someone whose grace and gentleness we would never forget.

We spent the rest of our time there filming her in the orphanage she had established years before and the love that flowed between this old lady and the dozens of children she cared for needed no words. Later I mentioned that I had never seen her wearing the Order of Merit, the UK's highest honour, which the Queen had presented to her. The award is a personal gift from the sovereign and, with only twenty-four people holding it at any one time, it has been described as the most exclusive club in the world. 'Come with me,' she said, leading me to her tiny private chapel. There, around the neck of a statue of the Madonna, was the medal. 'It is for her,' she said.

Mother Teresa died in 1997 at the age of eighty-one and in 2016 she was canonized as St Teresa of Kolkata.

After such an emotional encounter, could India offer anything else as memorable? If you worked for *Newsround* it could. How about coming face-to-face with a tiger? Our next destination was Ranthambore National Park in the state of Rajasthan and it involved a long train journey from Delhi. We hadn't eaten for many hours, so when a steward opened our compartment door and offered hot food he got a hero's reception.

'I only have chicken,' he said. 'Fantastic. That will be wonderful. Anything,' we chorused. But when the food arrived, we immediately lost our appetite. He produced four plates on which were lots of bones and a few traces of green, foul-smelling flesh. Maybe it was spices that were to blame, but so far on the trip we had avoided 'Delhi belly' and we didn't want to risk it now. So, hungry as we were, we emptied the plates onto the track through the large hole in the floor of the carriage's toilet (usually used for other purposes).

The vultures must have had a treat, and when the steward returned to collect the plates he looked at them in complete disbelief. We could guess what was going through his mind. *These strange Englishmen have eaten everything – even the bones!* We smiled and thanked him. 'Absolutely delicious,' we lied.

We were met at Sawai Madhopur, the nearest station to Ranthambore, by another legend of the subcontinent, Fatah Singh Rathore, known to everyone as the 'tiger guru'. An imposing man in his mid-forties, with a huge handlebar moustache, Fatah Singh almost single-handedly brought the plight of the tiger to the attention of the world.

In his open-topped jeep on the way to the reserve he told me that his first job, when he was working as a ranger for the Maharajah of Udaipur in the early 1960s, was to organize a tiger hunt for the Duke of Edinburgh. In fact the first dead tiger he saw was shot by the Duke. Maybe that moment changed them both – the Duke later became president of the World Wide Fund for Nature and Fatah Singh, who until then had been indifferent to tigers, became one of their main protectors.

Ten years later he became a wildlife warden at Ranthambore, which in days gone by had been the hunting grounds of the maharajahs of Jaipur, where guests would stay in a splendid lodge and ride out on elephants to 'bag a tiger or two'. Shortly before he took on the job, a census revealed that India's tiger numbers had dropped alarmingly to around 1,800. At the beginning of the twentieth century there had been perhaps 100,000.

Things happened quickly; tiger hunting was banned, Project Tiger was launched with the backing of India's prime minister, Indira Gandhi, and nine reserves were set up, including Ranthambore. When we arrived at the reserve, which was his pride and passion, I was overwhelmed by its size and beauty – 150 square miles of forest, hills and limpid lakes which in those days seemed quite untouched by human hand.

In the silence of this almost secret valley you could hear the calls of peacocks and langur monkeys. Eagles soared overhead and high on a hill was one of the few traces of human intervention – the crumbling ruins of an abandoned fortress, once the biggest in India but now in the possession of all manner of wildlife. We had entered a wilderness in its purest form.

Today it's very different, I'm told, because it has become a prime tourist attraction with luxury hotels and lodges. At the time of *Newsround*'s visit, apart from Fatah Singh and his team we were the only people there. 'Tracking tigers is much easier than people think,' said Fatah as we bumped along a forest track. 'All you have to do is drive around keeping a sharp lookout for fresh dung and when you spot it you then search for pug marks (paw marks) in that area. From the pug marks we can tell if a tiger is injured or is hungry, and usually they are not far away.

'We make sure conditions here are good for tigers but this is not a safari park. We do nothing to destroy the natural balance. If a tiger gets sick we do not try to cure it. We let nature take its course.

'Sometimes we go for days without seeing one but I have a feeling that perhaps you will be lucky.'

That certainly built up expectations as we settled down

for the night in the elegant remains of the Maharajah's old hunting lodge, with lacquered chamber pots still under the beds but only cold water dribbling from the taps. Next morning bright and early Fatah Singh was there in his jeep to take us on our first tiger patrol.

For some hours all we saw were exotic birds and sambar deer, but then Fatah suddenly slammed on the brakes – so suddenly that, standing in the back of the truck, I lost my balance and fell out. He shouted for me to quickly get back on board because there was danger around. He had spotted pug marks and then, just ahead of us, sprawled across the track, was a huge tiger. 'We call him Genghis Khan and you can see why,' said Fatah. 'He is big and very mean, just like the old Emperor, and he is very much the boss round here. We'll have to wait a while until he decides to move.'

'He can take as long as he likes,' said Ian our cameraman, 'I never dreamt I'd be filming a wild tiger so close.' Genghis Khan eventually sloped off into the undergrowth and we continued on our patrol, only to find another tiger slumbering in the shade of some bushes. We positioned ourselves so that I could talk to the camera with the tiger behind me and as I started to speak he raised his head and looked around. We couldn't have hoped for anything better.

To create this sanctuary for around forty tigers Fatah Singh had to make a difficult decision. Around a thousand people lived in Ranthambore in great poverty with no modern facilities. Some parts of the forest were being turned into desert by villagers and their cattle. He had the task of clearing sixteen villages and moving everyone and their livestock to homes outside the reserve.

'They didn't want to leave because their families had lived

there for many years but we could offer them a better life while creating a much more suitable habitat for tigers,' he told me. 'But I did weep with them because I knew the price they were paying.' The displaced families received compensation and new opportunities in a specially built town which had a health centre, school and good farming land and the places they vacated began to regenerate, slowly returning to forest.

The next day the tigers kept undercover and there were no sightings. That evening we moved to a small lodge on the banks of the tranquil Padam Talao lake and as we sat watching the sun go down Fatah Singh told us more about Genghis Khan, his favourite. 'One night something extraordinary happened,' he told us. 'I was sitting just about here and on the other side of the lake watched Genghis Khan kill a young sambar deer that had been drinking at the water's edge. He was standing back proudly, pleased with himself, when suddenly there was a splashing and a thrashing and the deer vanished. It had been snatched by a marsh crocodile – Genghis Khan's supper had been stolen from under his nose. That night he was a very hungry, very angry tiger.'

Fatah was the perfect guide to the kingdom he had helped create and, knowing that we had only a few days in which to film (unlike wildlife documentary film-makers who often have months, even years to get their shots), he certainly came up with the goods. He had been made field director of the reserve and was fearless in his protection of the animals he had come to love – he'd been beaten almost to death by poachers and was suspended for a while after catching a top government official illegally shooting wild boar.

Some years after my visit he reported that twenty of his forty-five tigers had vanished, presumably killed by poachers.

Ever since the hunting ban, the value of tigers killed for their skins and their body parts for use in oriental medicine had soared.

In the 1990s Fatah set up Tiger Watch in Ranthambore with a mission to protect not only tigers but also many of the local people living around the edges of the reserve. Throughout his career he battled with both poachers and the forest authorities who seemed to be unconcerned by the threat to Indian's most famous animal. 'I will never give up because, if I and others who believe as I do fail in our duty, there will soon come a day when this country has no more tigers,' he said. 'We cannot let that happen.'

Fatah Singh Rathore, the Tiger Man of India, died in 2011. In their different ways, he and Mother Teresa threw the spotlight on near-insurmountable problems and the world benefitted from their presence.

13. Zimbabwe-Rhodesia

Arriving at Johannesburg's international airport in the late summer of 1979 I was astonished to discover that I was banned from entering South Africa. As part of a series called *Newsround Africa* I was on my way to what was then Salisbury, the capital of the country which for a few brief months had the name of Zimbabwe-Rhodesia. I had hoped to do a quick 'recce' in Johannesburg. But when I approached the immigration control desk an official took my passport and then opened a rather thick book.

Going down the long list of names he must have come across mine because he said in a strong Boer accent: 'I'm sorry, Mr Craven, but you are not allowed into this country.' No further explanation was forthcoming, just a barrier that would not rise.

There had been no hint of any potential problems before leaving London but later I learned that BBC bulletins were regularly monitored by the authorities. Apparently they did not like the fact that every time *Newsround* mentioned apartheid we told our audience in clear terms just what the word meant; that the white minority of people in South Africa controlled the lives of the black majority.

If that regular repetition had raised the hackles of the censors then I was proud. There was no option but to wait in the no-man's land of transit for many long hours for the next flight to Salisbury. Though I have had many scary moments on aeroplanes what was to follow was one of the most frightening.

As I boarded the South African Airways Boeing I overheard a fellow passenger comment: 'Obviously this one is expendable.' As you can imagine, that did nothing to calm my nerves because, like everyone else on the plane, I was all too aware that the journey ahead was potentially one of the most dangerous in the world.

In the previous few months two Air Rhodesia Vickers Viscounts had been shot down in Rhodesian airspace by Soviet-built ground-to-air missiles in the hands of guerrillas (or freedom fighters, depending on your point of view) demanding independence. Thirty-eight of the fifty-six passengers and crew on the first plane died in the crash but to add to the horror ten survivors were rounded up and massacred by the rebels when they arrived at the scene. Eight people lived, three by hiding in the bush and the others because they had gone to look for water.

The second plane failed to get above the height needed to escape another of the shoulder-launched anti-aircraft missiles and crashed into rough ground near Lake Kariba after it was hit. All fifty-six on board were killed. With the rebels still out there with their launchers this, I reflected, was not a good time to be flying to Salisbury.

Before we took off, the captain explained in the reassuring manner they all have that we would be taking some precautions on the flight. It was night-time as we crossed the border into Zimbabwe-Rhodesia and window blinds were pulled down and lights switched off, so that we could not be spotted from the ground. We were warned we would put everyone in peril if we reached for our overhead light. So for 300 white-knuckle miles we flew in darkness and almost complete silence.

Then, as we approached Salisbury airport at 30,000 feet, the captain suddenly flung the plane towards the ground at such a speed and angle that we had to hold tight to the seat in front and my ears popped. It was like being on a fairground ride without the fun.

The captain was putting us through this in the hope that if the men with the missiles were waiting for us that night they would not be able to pinpoint their target. Who knows whether they were there or not, but our aircraft survived the descent and met the tarmac with tremendous force, jolting us in our seats. But who cared – we were still alive and everyone on board started clapping, cheering, praying out loud and crying. As we taxied to the terminal I told myself: *It's not over yet – you still have to get out of this country in one of these.*

It was a critical time to be in a land which until the 1960s had been (and here comes a *Newsround* backgrounder) named Southern Rhodesia and was a British colony. Many other African countries were gaining independence from their former European rulers at that time. But the white-minority government led by Ian Smith intended to cling to power and in 1964 issued a Unilateral Declaration of Independence (UDI) from the United Kingdom and created the Republic of Rhodesia.

This led to a brutal civil war in which 20,000 people were killed, with Ian Smith's forces on one side and black fighters of the Patriotic Front led by Joshua Nkomo and Robert Mugabe on the other. By the time my producer Eric Rowan (who was later to take over from Jill Roach as *Newsround*'s editor) and I began filming, Ian Smith had yielded to pressure and there was a new government run jointly by blacks and whites led by Bishop Abel Muzorewa, but the civil war continued.

Eric and I stayed at Meikles Hotel, which was then, and still is, I believe, one of Africa's great hotels. We got a cheap rate because the war had ruined the tourist trade and despite the sanctions imposed on the regime by almost every other country in the world, there seemed to be no shortage of the good things Meikles was famed for.

We met up with our crew, freelance cameraman François Marais and his sound recordist Carol Clark – a very brave woman indeed, given that she and François travelled almost every day into what was known as 'bandit country', where every white face was a target. They told me it was pointless trying to explain you were a BBC journalist and flashing your press card because the locals believed you were agents of the hated white leadership they were determined to overthrow and which, anyway, had issued the press card.

As we drove round Salisbury everything seemed peaceful and orderly, like a suburb back home apart from the jacaranda trees starting to flower everywhere. But the picture changed when we headed into Harare, which was then a large black township on the outskirts of the capital and named in honour of a nineteenth-century chief. It felt a dangerous place, a tinderbox where anything could happen. But François, who'd been working in the country throughout the civil war, knew the safe areas and found some local children who were happy to talk to us about their hopes for the future. After independence the name of the capital changed, of course, from Salisbury to Harare.

Later we went into the countryside and I was astonished at how beautiful it was. Because much of this landlocked country is around 900 metres above sea level the climate was subtropical and almost anything could grow in its fertile soil all the year round. It was rightly known as the breadbasket of Africa. Maize was the staple crop but there was also millet, wheat, barley, peanuts, cotton and tobacco. This truly was a land of plenty but most of it was owned by white farmers, and the black majority wanted them out and the farms for themselves.

As we drove through wooded valleys the landscape looked more like Scotland or the Lake District, but at any point the road could have been landmined and the trees on either side could be harbouring an ambush.

'One of the favourite tricks of the guerrillas is to put a herd of goats in the road to slow down a vehicle and then zap everyone in it,' said François, who was at the wheel. 'It's just one of the risks we have to take if we want to report this war properly.'

Only a few seconds after saying that we turned a corner of the weaving, single-track road and there in front of us, blocking our path, were some goats. Luckily not a whole herd, but my heart leapt as François sounded his horn to scatter them and put his foot down. Fortunately for us there was no gunfire – just a few loose animals.

On our way to meet a white farming family, he turned off the main road onto a rough track leading to their farmhouse and suddenly stopped the car. 'This does not look good,' he said. 'I drive on only tarmac roads, because there is less chance of landmines being planted. This is an unmade road and I don't like the look of it. We can turn round or we can take a chance, as there haven't been any attacks round here lately. So, if we all agree, let's take a chance.'

We did just that, with fingers crossed and prayers on our lips, and after a nightmare few miles arrived in one piece at the farm – only for the owner to tell us that he had just swept the road for mines with his armoured vehicle but had forgotten to tell us when we had called him from the hotel.

He was a good man who treated his black workers well but he, his wife and young daughters could well be next on the list for a visit from the terrorists/freedom fighters. He showed me the impenetrable 'safe room' in the centre of the house where his family would shelter while he would single-handedly attempt to save them and his property. I never discovered whether he was ever put to the test.

Back safely in Meikles, next morning's front-page headline in the *Herald* national newspaper was 'Mountbatten Killed'. The Queen's cousin had been blown up by terrorists while on a family fishing trip off the west coast of Ireland.

On that sunny morning in Salisbury it just confirmed my fears that the world seemed to be an increasingly dangerous place, from the starkly beautiful landscape of Ireland to the golden fields of southern Africa. Awful as the news was, it reminded me bizarrely of a conversation I'd had a few years earlier with a producer called Ron Webster, who was in charge of compiling major obituaries for BBC TV.

'Many of these obituaries are put together months even years ahead of a death and I asked Lord Mountbatten if he would contribute to one I was making about someone he knew well,' Ron told me. 'He agreed and was rather good. Then he said he liked the idea of writing and introducing his own obituary, which took me rather by surprise as no one had ever done that.' Mountbatten was a man who liked to create and control his own public image.

Also on that front page, Robert Mugabe was expressing serious doubts that the forthcoming talks in London aimed at ending the war and setting up an independent Zimbabwe would be successful. During our stay we talked to white children in bombproof shelters and black children in their deprived villages and came away wondering what would happen to them all once this wind of change had blown through their country. I couldn't feel too positive about the sort of freedom they might face. My fears were confirmed on another visit some years later when one of the team at the state television studios confided to me: 'We can't begin the news bulletin until the scripts arrive from the government office.'

On our last night, there was a dinner dance at Meikles and most of the guests were white farmers and their wives. The atmosphere was more like London in the late 1940s than Africa in the late 1970s; the orchestra played Glenn Miller and afterwards as the ladies collected their elegant coats from the cloakroom their husbands were handed back the automatic rifles and guns they had checked in. As they set off on the potentially deadly journey home they were playing one of the final scenes in British colonial history.

You could sense that all around an era was ending for both the until-now privileged whites and the newly powerful blacks. The talks at Lancaster House in London the following month brought independence and Robert Mugabe won the battle for control over his rival Joshua Nkomo and stayed all-powerful until his overthrow in 2017. Many white farmers were forced to leave as their lands were taken over by the state and, in sharp contrast to its breadbasket days, Zimbabwe has been in pretty poor shape ever since.

Our airliner did a steep spiral around Salisbury before reaching a safe cruising height away from all those missile launchers. Safely back in Johannesburg, immigration checks were obviously not as stringent between the two African nations because we were allowed into the city 'for shopping and sightseeing only' for a few hours before catching the London flight.

And there were no border problems a few years later when Marilyn and I stopped over in Durban on our way home from a cruise. That evening during dinner we got into conversation with an English couple who, it turned out, had emigrated to South Africa. 'What made you leave England?' I asked him, expecting to hear about bonuses like lots of sunshine and tax benefits.

'Esther Ranzten!' came the blunt reply. Apparently he'd been running a dodgy company that had come under the scrutiny of her *That's Life* programme and had had to beat a hasty retreat to the more understanding surroundings of Durban.

After I returned from Zimbabwe-Rhodesia the distinguished novelist Margaret Forster wrote about *Newsround* in the *London Evening Standard*: 'I cannot do without it. The trouble in Rhodesia was explained this week succinctly, with no punches pulled, and it's the first time I've ever really known what stage the whole conflict has gone to. My, one does learn a lot from children's telly.'

14. The Falklands

March 18, 1982. It was a quiet news day: the first space shuttle, *Columbia*, was being prepared for its maiden launch but apart from that not a lot was happening in the world that might interest the *Newsround* audience. Then we spotted an obscure item on the news agency tapes about a gang of scrap-metal workers from Argentina who had landed on a remote British island in the South Atlantic Ocean and for some reason hoisted their country's flag.

What was that all about? The island in question was inhospitable, mountainous, unpopulated South Georgia, a place mostly covered in snow and ice and home to huge numbers of seals, penguins and seabirds. The famous polar explorer Sir Ernest Shackleton is buried there.

The scrap men said they intended to dismantle an abandoned whaling station, but were they really who they claimed to be? Could they be an undercover unit of the Argentinian armed forces? After all, it was common knowledge that Argentina claimed sovereignty of South Georgia, just as it did of the Falkland Islands, which were around 1,000 miles away to the north-west. Is that why they'd raised the flag?

This had all the ingredients of a *Newsround* backgrounder – a place we'd never heard of close to Antarctica, a hero explorer, great wildlife and mysterious happenings. And there was also drama – our government was ordering these men to pull down their flag and leave. We had our offbeat story of the day, but little did we realize that this minor incident would lead to the biggest UK military involvement since the Second World War – the Falklands Conflict.

As events unfolded it became clear that some of the scrap-metal team were indeed Argentinian marines, and though they did agree to abandon their mission a much larger force returned almost two weeks later to capture the island. And an even bigger force invaded the Falklands.

That happened on a Friday and, as *Newsround* was off the air, we were busy planning a little celebration for our tenth birthday the following Monday. Ten children born on the same day, 4 April 1972, had been invited to a lunchtime party in London and we were busy filming the fun when a dramatic message came through from our office that the invasion had started and we would be reporting it in a special, unscheduled edition.

Fortunately we had completed much of our filming so, making my excuses and leaving the children to their buns and balloons, I dashed back to Television Centre. Because *Newsround* had been across this story from the very start thanks to the 'scrap-metal saga', the schedulers found two or three minutes of airtime that afternoon in which we explained to our viewers that the UK was not going to allow Argentina to take over some small islands half the world away that had been British since 1840.

Next morning the House of Commons had an almost unheard-of Saturday session in which Mrs Thatcher told MPs that the Falkland's tiny capital, Port Stanley, had been taken and the governor, Sir Rex Hunt, had flown to Uruguay on the South American mainland.

The events that followed – the Task Force sailing south, the invasion and the eventual recapture of the islands, the stories of heroism and tragedy – filled our bulletins for the next couple of months. On air, we carefully explained the political manoeuvrings and the price that was being paid in lives, and (some readers might find this remarkable) there was never any pressure from any quarter for us to toe any political line. In fact that never happened at any point during my days on *Newsround* – it was solely our responsibility to be fair and impartial. No one outside our little team ever told me what to say.

When HMS *Sheffield* was hit by an Exocet missile and twenty crewmen died I happened to be filming in the playground of my old primary school in Leeds. Looking around the young faces, I was reminded forcefully of just how careful we had to be in reporting all aspects of the confrontation.

Any one of those faces could have a father or other relative in the front line. On the day I reported that the conflict was over and the Falklands were back under British rule we ended *Newsround* with footage of the funerals in a mass grave of the UK soldiers killed at Goose Green, the first major battle, as a piper played a lament. The average delay between material being filmed in the Falklands and shown on UK television was seventeen days. All news coverage had to be sent home by ship and plane because the military said they needed all satellite bandwidth. So those poignant images from Goose Green had arrived on the same morning that British troops marched into Port Stanley and the Argentinians surrendered. Without over-egging the message, it was a fitting example of the human cost of any conflict.

In the aftermath I joined BBC outside broadcast teams at Portsmouth as they covered live the joyous homecomings of the Task Force. First Royal Naval vessel to arrive was the flagship HMS *Hermes* and I was on the quayside making my way through expectant, cheering crowds of relatives as they tried to spot their loved ones lining the decks of the huge warship. It was a carnival-like occasion with huge welcoming banners and seas of Union flags and the scenes were repeated a few weeks later when her sister aircraft carrier, HMS *Invincible*, returned with the rest of the Task Force.

I did lots of 'vox pops', quick chats with anyone who was willing to talk to me, and on live television people can say the most unexpected things. One little boy told me: 'My mum just can't wait to get my dad upstairs again.' What on earth would he say next, I wondered, fearing the worst, but before I could stop him he added: 'Because he's only half-finished painting the bathroom.'

Sighs of relief all round. Later I was talking to one family when an anxious-looking woman pushed her way through. 'Are you looking for someone?' I asked her. 'No,' she said, 'I'm looking for the loos.' I told her I thought I'd seen some further along the quay. She smiled gratefully and I carried on with my interviews.

That October I had the honour of being guest speaker on board the training ship HMS *Kent* in Portsmouth on Trafalgar Night, when every year the Royal Navy celebrates Nelson's great victory of 1805. This was a particularly poignant night because some young officers who had lost their ships from under them during the more recent sea battle had been posted to the *Kent* while they awaited new orders. Not a great deal was said, at least not to me, about their experiences but it was a spirited evening which ended in the customary way with some rough-and-tumble mess games.

The cruise ship SS *Uganda* had been seeing out her days taking children on educational voyages around the coasts of Europe when she was rapidly converted into the Task Force's hospital ship. The work was done in Gibraltar in six days and I was there for *Newsround* to watch her leave for the South Atlantic where she would fulfil a very different and unexpected role. She treated more than 700 casualties and one of her officers told me Argentinian fighter planes flew past below her bridge line but respected her huge Red Cross markings. On her return the *Uganda* went back to her old and happier job and on her first sailing I was on board with a *Newsround* film crew as she made her way down the English Channel cheered on by hundreds of excited young voices. She was scrapped in 1992.

My final connection with the Falklands in 1982 came on New Year's Eve when I was hosting BBC One's coverage of the festivities from St Katharine Docks on the Thames. At about ten to midnight I was due to talk by satellite link to Sir Rex Hunt, long since restored as the governor of the islands. The idea was to have a conversation about the mood of Falklanders, after what they had been through, as the new year approached. But our engineers could not establish contact with Sir Rex so we continued without him until, at about a minute to twelve, I got word in my ear that he was finally on the line.

There was no time for a proper chat because, as the whole nation knows, a few seconds of anticipatory silence must be observed as the great clock works its way up to the midnight bongs. We exchanged a few brief words and I wished him and all Falklanders a happy and peaceful new year and he reciprocated. Just a few seconds that ended a momentous year in the South Atlantic – then cue Big Ben.

15. Two Early Stars

Children's programmes have a proud history and it all started with two terrific people – one a First World War fighter-pilot hero, the other the BBC's first woman announcer – who were both in their eighties when I was lucky enough to meet them for a television documentary I was making called *In Front of the Children*. These voices from the past, from the sepia days of broadcasting, were the haphazard pioneers of a tea-time habit which is still with us nearly a hundred years later.

In 1922 Cecil Lewis and Kathleen Garscadden picked up their microphones, he in London and she in Glasgow, at five in the afternoon and said hello to whoever might be listening to this revolutionary medium they called the wireless – and *Children's Hour* was born. Somewhere in the wilds of Perthshire a little boy was spellbound as his new-found radio 'aunty' told a story about a butterfly – and suddenly, to his jaw-dropping amazement, the real thing fluttered out from the great horn-shaped loudspeaker. 'He wrote thanking me for sending the butterfly,' Kathleen told me. 'He really believed it, and who could blame him. The wireless was such a wonderful innovation . . . hearing voices through the air.' The magic had begun.

Cecil Lewis was only eighteen when he won the MC as a pilot above the battlefields of France and Belgium, and four years later he joined the British Broadcasting Company (as it then was) as deputy programme director. Impressive, but then there were only four people on the staff. On *Children's Hour* he was Uncle Caractacus – 'Uncle Cecil would have been too awful, too sissy' – getting 150 letters a week, yet the great John Reith, founding father of the BBC, didn't even know the programme was being transmitted.

'He paid no attention whatsoever,' Cecil told me during a brief visit to London from his home in Corfu. 'He was entirely occupied with serious matters like getting the Post Office to pay up our share of the licence fee. In fact, he didn't bother about programmes at all, ever.'

Cecil, on the other hand, cared passionately about them. He was one of the most engaging people I've ever met and his romantic, adventurous life took him from China to the South Seas to Hollywood. It felt an honour to be in his presence – and what a presence: 6 feet 4 inches tall, boundless enthusiasm, an infectious laugh and a fount of tremendous reminiscences from a glittering career.

He was back and forth at the BBC over many years and among his achievements was an Academy Award for his screenplay of *Pygmalion* (George Bernard Shaw would only allow it to be filmed if Cecil wrote the script). His bestselling memoir, *Sagittarius Rising*, about his wartime flying experiences became a classic – but among his fondest memories were those early days of broadcasting:

> *Children's Hour* had a direct and immediate appeal. It seemed to touch children – it was something that was simple and easy to get hold of. We were all in our twenties, pretty jolly and fooling around so there was a spontaneous cheerfulness about the programmes that was infectious. It was a big success – our first success.

Came five o'clock we thought we've done enough in the office for the day so we'd better start the children's hour. We clapped on our hats and ran down the Kingsway to the Gaiety Theatre, got into the lift, went up to the studio and waited for someone to say we were switched on. It was absolute improvisation from beginning to end.

We had a good and dear chap called Stanton Jefferies who improvised music, I told a story and someone else read out letters. It all began in a most haphazard way, with no sense of being planned or directed or important at all. Nothing was ever written down. We were trying to run the whole of the BBC with just four people so we didn't have time.

In Glasgow, Scotland's first radio star Kathleen Garscadden was enjoying the fun as well. Still bright as a button when I met her, with a personality as warm as her lilting voice, as we walked along the banks of Loch Lomond where she used to take some of her young listeners on picnics she told me:

> I was just a youngster called in to speak at the microphone because no one else would do it. People were terrified of microphones. We sang in a bunch – engineers, office people and commissionaires – anybody who could be lured into a studio. It was just a lovely party all the time. We gave each other silly names. To begin with I was Aunty Cyclone and someone else was Aunty Macassar. When the British Broadcasting Corporation came along they didn't approve of Aunty Cyclone. They said it would confuse the children, they wouldn't know the difference between me and the weather forecast, and so I had to become Aunty Kathleen.

Young listeners with their primitive 'cat's whiskers' receivers were linked together by the Radio Circle, the first successful attempt at audience participation. As radio stations sprouted around the country each had its own circle with membership costing a shilling a year, later reduced to ninepence, which went to charity. For that they got a certificate, a calendar and a badge with a pendant bearing the name of their station.

Eighty thousand children joined, and from the start they were helping others, collecting silver paper to pay for cots in local hospitals. As one writer remarked: 'Long after children who belonged to the Radio Circle have grown up these cots will remain as a reminder of the days when broadcasting itself was young.'

After ten years the Radio Circle was killed off, the victim of its own success because all the members expected their names to be read out on air on their birthdays. Said Aunty Kathleen: 'Eventually there would have been no time for programmes – just birthday messages. But they were fun while they lasted. A relative would write in saying, "If wee Willy looks under the coal shovel he'll find a present," or, "Go under the bath and you'll find a treat from Granny."'

The *Radio Times* reported, upon the Radio Circle's demise in 1933: 'On one particular occasion the number of names to be read out was 572, taking up twenty-two minutes. The broadcast radio greeting was a relic of the old, informal days and the recipients' pleasure in hearing their own names read out compensated them for the tedium of hearing so many other people's.'

So, Aunty Kathleen told me, they involved the audience in other ways with concerts, outings and Christmas parties, and there was even a *Children's Radio Magazine*. But BBC bosses frowned on the happy-go-lucky style and ordered that the programmes should be better organized.

Some even wanted them scrapped, but they stayed for more than forty years, until 1966 by which time they had lost the audience to television.

One of the last radio uncles, David Davis (how I remember as a child listening to his soft warm voice) had worked for *Children's Hour* since 1935 and had introduced Princess Elizabeth when she made her first appearance on the programme in 1940. Its philosophy ran through his veins and he told me: 'When the end came it was one of the saddest days of my life. I took the last programme myself and, rightly or wrongly, chose Oscar Wilde's "The Selfish Giant" as the story to end on. It was a slightly silly crack at the BBC and rather naughty but I felt that a great giant at the BBC had closed its jaws on a very precious thing, just like the giant in the story.'

Uncle Caractacus, Aunty Kathleen and Uncle David, who all died in the 1990s, were themselves among the friendly giants of broadcasting and all of us who followed in their steps owe them a great debt for they established the rights of children everywhere to be educated, entertained and heard across the airwaves.

16. Hmong

It was purely by chance that I came across the tribe that changed continents – and it all happened because I was covering a space mission. From its jungle base at Kourou, 300 miles north of the equator in the South American country of French Guiana, the European Space Agency has been launching rockets carrying satellites into orbit for more than forty years.

Newsround was there in the mid-1980s to watch the launch of an *Ariane* rocket as part of a programme we were making about space technology. Our team was allowed onto the launch pad just hours before the countdown and I was standing a few inches from the giant tube filled with the fuel that would blast it off from Earth. Now I do have a tendency to lean on things, and the chief engineer suddenly shouted out 'Don't!' He thought I was about to relax against the side of the rocket.

'The skin is only millimetres thick,' he said, 'and if you touch it you will dent it. Then we would have to cancel the launch at a cost of millions.' I tried to reassure him that on this occasion I had no intention of leaning but he wasn't taking any chances and ushered us away from the platform. The launch that night was spectacular. Ninety per cent of French Guiana is tropical rainforest and our observation area was in a clearing some miles from Kourou; all around was dark and silent.

Suddenly there was a blinding flash, a great roar and the ground seemed to shake as the *Ariane* rocket hurtled towards space. The noise lasted for minutes and then silence returned, but only for a short time. *Ariane*'s deafening blast-off was replaced by a cacophony of animal and bird sounds as the creatures of the jungle, rudely awaked, seemed to be saying to each other: 'What the heck was that?'

The next day we were taking shots of the surrounding area and stopped for lunch at a primitive café along a jungle track. But, Guiana being French, the only hot food on offer was coq au vin. It was delicious and when I asked where they kept the chickens I was told: 'No, we cook birds of the forest'. Only later did we discover that we had been eating parrot.

During our meal a group of tough-looking guys walked in and ordered beers. They were obviously off-duty French Foreign Legionnaires – the Legion did its tropical training in these parts – and one of them stared at me in a somewhat menacing way and began discussing us with his pals. Then he got up and walked slowly towards us. It looked as though we might, for no good reason, be getting into a fight. But when he was really up close to me he said: 'What are you doing 'ere, John? Are you filming for *Newsround*?'

He was from High Wycombe and had joined up because he was bored and unemployed. One morning he had been reading about the Legion in the *Daily Mirror* and the next morning he was in Paris signing away his life for the next few years. Now he had his regrets, but because he made the mistake of handing over his passport at the recruiting office he could not leave. Many Legionnaires are escaping from the law and give

22. Interviewing Pelé in New York.

23. In Thailand getting close to an elephant for *Newsround Extra*.

24. *Multi-Coloured Swap Shop*, the early days.

25. The *Swap Shop* team.

26. Emma after the accident. Her teddy also needed a bandage and inadvertently started a trend . . .

27. Winning the *TV Times* award for Children's Favourite Presenter.

28. Getting gunged on *Crackerjack*. The things we had to do!

29. Interviewing Prince Philip about the Duke of Edinburgh Awards. No prizes for my flares!

30. Going ashore after filming on school ship SS *Uganda* in the English Channel, 1982.

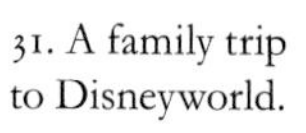

31. A family trip to Disneyworld.

32. Hamming it up with Paul Daniels and Joanna Lumley at Smithfield Market, 1980s.

33. Cubs tea party on the Thames – and one of my favourite jumpers.

34. Marilyn and me at a Eurovision conference, Athens.

35. My *Newsround* 'mug-shot' sent to viewers who'd asked for a signed photo.

36. At the *Newsround* studio desk with Helen Rollason and Roger Finn.

37. Team shot with studio crew on *Newsround*'s 15th birthday, 1987.

38. Being inducted into the RTS Hall of Fame, 1996.

39. At Buckingham Palace with daughters Victoria (*left*) and Emma (*right*) after recieving the OBE for services to rural and children's broadcasting, 2000.

40. My very rural 25th anniversary cake from the *Countryfile* team, 2014.

41. Filming an episode of *Countryfile Diaries*, 2018.

false names and no passports. My new friend had been rather innocent at the time but had certainly toughened up.

After the successful launch we were looking for another story to film in the area before our journey home and one of the engineers said: 'Why don't you film the tribe that lives just down the river from here? They have an interesting story to tell.' Intrigued, I asked him the name of the tribe. 'They're called the Hmong,' he said. 'They can't be,' I replied, incredulously. 'The Hmong are a wandering hill people from South East Asia. I filmed with them some years before in a refugee camp in Thailand.' 'That's the same tribe,' said our engineering friend. 'They were shipped here in jumbo jets.'

So I had stumbled across an exclusive story that would later be shown around the world. It started seven years earlier on the banks of the Mekong river, which separates Thailand from Laos. Thousands of people were fleeing from a revolution in Laos – not as brutal as the one in neighbouring Cambodia, where a third of the population was butchered on the orders of the dictator Pol Pot, but bad enough to make many fear for their lives.

About 30,000 had swum or rowed across the Mekong and some had ended up in the temporary camp at Non Kai. We had great trouble getting permission to film in the camp because the local police chief demanded a large bribe which we were not prepared to pay. We had dinner with him in a ramshackle cafe near the camp and his mood became more and more aggressive. Then, without warning, he picked up a raw egg from a bowl of them on the table and cracked it into the almost full lager glass of our cameraman, Oggie Lomas, and glared defiantly at him.

Totally unfazed, Oggie slowly downed his drink, egg and

all, and, smiling at the police chief, did the same to his lager – in went an egg. Now the challenge was on, round after round, until our unpleasant policeman slid beneath the table, still clasping a glass of egg and lager. 'That was fun,' said Oggie later. 'Don't you feel ill?' I asked him. 'Not at all,' replied Oggie, 'I have raw eggs for breakfast every day!' Next morning the shamed police chief reluctantly issued our filming permit.

Some of the refugees we saw in sprawling Non Kai were nomadic Hmong tribespeople who had escaped from the mist-shrouded uplands of Laos. They had been allies of the US and were abandoned when the Americans pulled out. As I reported then:

> For the first time in their lives they had been herded into one place – the camp – and they were not allowed to move from it. They looked completely dejected, walking aimlessly around their compound, knowing that they would never be able to return to their hills.
>
> It was as if a huge cloud of hopelessness hung over them and drifting along in the air was the plaintive sound of their music. It came from strange-looking wind instruments, their own version of the pipes of Pan. Some of the children showed me how to play the pipes but only they could capture the wistful sound that reminded them of their lost land. What on earth will happen to these people? Will they be here for ever? They have never known life in towns and cities so where can they go?

Now, I had found my answer 10,000 miles away in a country at the top of South America. The French authorities, who had previously ruled Laos, arranged for the Hmong in the Non Kai camp to be flown lock, stock and barrel to a jungle wilderness on a different continent – and, as I was to find out, they were thriving.

Our team borrowed a boat and headed downriver from the space-age surroundings of Kourou into the unbroken rainforest until we rounded a bend and came across a clearing. In it were several dozen Laos-style houses on stilts and, as we got closer, I could see men, women and children with the distinctive appearance of the Hmong. It was a totally unreal moment – to find an Asian village in a South American landscape. We tied up the boat and a group of villagers approached us looking distinctly wary.

Then an amazing thing happened – two of the elders came right up to me, examined me closely and then relaxed and began to smile. They remembered me – and I them – from the camp. A young man was summoned who spoke passable English and I explained that engineers at Kourou had told me about them and I had come to see them in their new home and wish them well. Once that had been translated, we became honoured guests.

The elders proudly showed us round the village, which they called Cacao, and were more than happy for us to film their new lives. They told us they had no idea what to expect when they saw airliners for the first time and were astonished when they stepped out from them into an entirely unknown environment. But they liked the land they were given, and the climate, and were making the most of it.

Back in Laos, they grew food for only their own families but in French Guiana they needed money to live so they developed their agricultural skills and carved farms out of the rainforest. They sold their fruit and vegetables at the weekly market in the capital, Cayenne. The women were keeping alive the old traditions of dancing and embroidery – even making tapestries depicting their exodus from Laos – and passing on their memories to the children, who were having a great time splashing around in the river.

I even heard some of them playing their wonderful pipes. We spent a very happy, unexpected day with the Hmong and when I got home their story made a fascinating *Newsround Extra* and I was also asked to compile a special report for the main evening news.

That was thirty years ago now, but this experiment in cross-continental tribal rehousing seems to have worked in the longer term. Though the Hmong make up only 1 per cent of French Guiana's population they now produce around 70 per cent of the country's agricultural output. They are doing well – those wooden huts have satellite dishes and the latest technology. I'm told that some of the elders would like to return to Laos, but just for a visit, while the youngsters don't seem interested.

At first some locals resented the newcomers but any tension seems to have receded. There are concerns, though. Because the 2,000 plus villagers don't mingle much with outsiders there is the danger of inbreeding, and with a high work ethic too many children are labouring in the fields and not the classrooms. But by and large the tribe that switched continents seems to have made the right move.

17. Death of a President

Over the years I've had the pleasure (well, usually it was a pleasure) of filming with many leading figures, from political leaders to pop stars, and nearly always in circumstances carefully managed by their 'people'. One of the most memorable happened in an African country I'd barely heard of, with a capital city whose name I had no idea how to pronounce and a leader I knew nothing about – respectively, Burkina Faso, Ouagadougou (Waga-do-goo) and President Thomas Sankara.

As one of my distinguished newsreader colleagues on the grown-up news once said to me: 'The trouble with these leaders in emergent African countries is that no sooner do you learn to pronounce their names correctly than they get bumped off.' And that was certainly true in the case of President Sankara. The only British television news bulletin to report his assassination on 15 October 1987 in any detail was *Newsround.* Other programmes, preoccupied with the storms in southern England and the crisis in the Gulf merely noted the death of one of the developing world's most original and idiosyncratic young leaders.

But *Newsround* laid him to rest with honours and he was especially mourned by two young viewers who, only a month before, had interviewed him. Dan Meigh, aged fourteen, from the Wirral and eleven-year-old Becky Branford from Brixton had won a competition organized by Sport Aid and

Newsround to find school-age journalistic talent – and their prize was to make a documentary film for children's TV about aid projects in the impoverished West African nation of Burkina Faso, which translates as Land of Dignified People. Other young winners filmed in Zimbabwe, the Sudan, the Far East and South America.

Thomas Sankara had changed the country's name from Upper Volta, and that wasn't the only thing he changed. In rundown Ouagadougou Dan and Becky interviewed him in his presidential palace. For security reasons we were not allowed to film the actual building or the wide avenue leading to it – the avenue down which the thirty-seven-year-old Sankara jogged twice a week with all his staff and Cabinet in tow, by presidential decree. No one in the palace escaped compulsory physical exercise.

It was a palace in name only, a large and undistinguished house left over from French colonial days, badly in need of new paint and dusters. Sankara could not be accused of squandering money on the place, nor on his transport. Parked outside was his official car, a Peugeot 205. He was not, as you will have already gathered, your standard African dictator. Soon after seizing power in August 1983 he ordered his ministers to surrender their fleet of Mercedes limousines and offered the vehicles as prizes in a national lottery. It was one of his characteristic moves as he campaigned against corruption and you don't do that kind of thing in that part of the world without making powerful enemies.

Dan and Becky were amused by the two ceremonial guards at the front door, garish in their orange battledress, slovenly in the heat, like characters from a comic opera. But their machine guns were real enough and they may have been among the few who tried to save Sankara when he was betrayed and killed by his best friend. After Captain Blaise Compaoré, who had been Sankara's closest adviser, led the coup and stormed the palace, a dozen of the president's men lay dead alongside him.

The only clue we had that trouble might be ahead was that 'Tommy', as Dan and Becky liked to call him, had not attended the summit of French-speaking heads of state in Canada. President Jean-Baptiste Bagaza of Burundi had been there and was overthrown in his absence. Sankara obviously thought it would be prudent to stay at home, but had he made the trip he might still be alive today.

We were shown into a dingy office, apparently the Cabinet room, where the *chef de protocol* had insisted the meeting be held. Great fuss was made of the children as they calmly prepared their questions for what was obviously regarded by the advisers as a major interview. No, they wouldn't keep him more than thirty minutes. Yes, they were being well looked after and were impressed by the UNICEF aid schemes out in the countryside in one of the world's poorest countries.

After a long wait Tommy appeared in a sleeveless blue kaftan-style robe, looking remarkably like Lionel Richie. Perhaps the image was reinforced by his reputation as a rock fan and rather good guitar player. By coincidence, both Dan and Becky were clarinettists and had brought along their instruments in the hope of playing with the president. I didn't try to deter them – what a little scoop it would have been, two British schoolchildren having an impromptu jam session with a flamboyant African dictator.

Sadly, it did not happen. Though Sankara was keen, the *chef de protocol* ushered him away to his next appointment. Looking over his shoulder, he smiled at our young newshounds and said: 'People talk about human rights. What about presidents' rights?'

During the interview he had been charming, forthright and open, agreeing to make his most important replies in English while protesting that too many of his teachers were in earshot to hear his grammatical mistakes. The questions were simple and, like all those from children, to the point. He admitted there was much to do if his country were ever to escape from poverty and said his ambitions were to improve health and education. His government had set up primary health-care centres manned by trained volunteers in more than 7,000 villages.

His basic Marxist beliefs were strong but so was that freewheeling, verging on zany, philosophy that almost certainly cost him his life. This was the man who sacked his government every July and sent them out to work in the fields. Each September they discovered whether they were part of the new administration or still out in the wilderness beyond Ouagadougou. This was the president who hitch-hiked home from a meeting of African heads of state to save money.

During the Race Against Time in 1986, when 20 million people in eighty-nine countries simultaneously took part in a ten-kilometre fun run to raise money for aid projects in Africa, Sankara ordered all his government workers to join in and led them himself. Television pictures of that were shown worldwide, ensuring that, for one day at least, his country was firmly on the map.

In his conversation with Dan and Becky, Sankara spoke

passionately about his ambitious, near impossible plan to create a new green belt across northern Burkina Faso, which was rapidly becoming part of the ever-growing Sahel desert that was encroaching at the rate of seven miles a year. He wanted to send 100,000 people out there to plant trees using money from the aid agencies. Already he had experience of mobilizing vast numbers to work on construction projects and now he wanted to turn the desert green.

'In your country you have candles on a cake to celebrate your birthday,' he said to the children. 'In my country I want people to plant a tree on their birthdays. That could be eight million trees a year.' More pragmatically he wanted to find alternative sources of energy so that trees would not have to be felled to use as fuel for cooking. Every year, he said, the number of carts in Burkina Faso carrying firewood could stretch the length of Africa from Cairo to the Cape.

We spent one week in his country and less than one hour in his company. Travelling around we saw scant sign of repression – just overwhelming poverty. Those who killed him claimed he had betrayed the revolution and sold out to the capitalists. Maybe he had, but we were not there long enough to make any judgement.

Tommy certainly made our newshounds feel like important people. Dan asked him what he would do with the money allocated to Burkina Faso from the Race Against Time. 'The first thing,' said the president, 'will be to buy a stamp and write a letter saying thank you.' The children wanted to ask him whether he felt in danger but the thirty minutes were up so they ended by asking him if he would run in the next Race Against Time. 'Only if you will sponsor me,' he said with a big smile. 'Ten pence a mile,' said Dan. 'Five pence from me,' said Becky.

'Then I will,' said President Thomas Sankara, and he would have done. He was one of the dignified people, and I can't help feeling that West Africa would today have been a gentler place had he lived.

18. Pandas

Since day one animal stories have been a mainstay of *Newsround* for the very simple reason that children love them – many young viewers have pets of their own and are fascinated by all types of animals, especially those in danger of extinction.

During my time on the programme we would try for at least one animal story in every edition. As I mentioned earlier, the lead on the very first programme on 4 April 1972 was about rare ospreys returning to a nest at Loch Garten in the Cairngorms, with an explanation of how these magnificent birds of prey had been persecuted for many years but were now protected and beginning to thrive. Since then some have even taken up residence in England, at Bassenthwaite in the Lake District, where I had the honour of officially opening the osprey observation centre in 2003, and on Rutland Water – the first ospreys to nest in that area for 150 years.

For *Newsround* the giant panda became a powerful symbol of the need to protect the world's most vulnerable creatures. Once, giant pandas thrived right across China and other parts of Asia, but by the 1970s they were on the red list of endangered species with perhaps 1,200 clinging on in the wild. I like to tell myself that one reason children in the UK have become more aware of and involved in wildlife issues over the past few decades is *Newsround*'s panda stories.

Not all of them have been serious – we have had some panda fun as well. One April Fool's Day in the 1970s I put in a request to the graphics team for a very large white egg (an ostrich egg, I think) and we painted black patches on it. Then, with the co-operation of London Zoo, it was gently placed in the panda enclosure – at that time the zoo's two pandas, Chia Chia and Ching Ching, were its biggest stars.

We filmed them peering at this strange object, even nudging it, and on that night's *Newsround* I excitedly announced that a panda egg had been laid for the first time, but it would take exactly a year to hatch, so both zoo officials and the pandas would have to wait until 1 April next year to see Britain's first baby panda. It's amazing how many viewers were fooled, and not just children. Several letters came in, and there were calls to the duty office as well, angrily pointing out that we had misinformed our audience, and what kind of journalists were we if we believed that pandas laid eggs.

Of course, we on *Newsround* weren't the only ones to capitalize on the extraordinary appeal of these black-and-white bears. I was told that at an inaugural meeting of the World Wildlife Fund in 1961 one of the founders, the renowned naturalist and broadcaster Sir Peter Scott, was doodling on his notepad during some of the more bureaucratic discussions. When the question arose of what should be the symbol of the WWF, the person sitting next to him said: 'I think Peter has just designed it.' He had been doodling a panda – and that little drawing remains the WWF symbol to this day.

WWF sounded the alarm when, in the early 1980s, the panda's staple food – arrow bamboo – began to flower across vast stretches of their habitat in the Sichuan province of south-western China. When that happens, the plant withers and dies, and it can take up to ten years for new shoots to grow. The tragic result was that several hundred pandas starved to death and a rescue operation was mounted by the Chinese authorities and WWF to try and save others from the same fate.

Because *Newsround* had established a reputation for its panda coverage, I was invited to be the first Western journalist to film that rescue mission. In 1985 I flew to Chengdu, the capital of Sichuan, a province which is one of the last strongholds of the panda, with director Nick Heathcote and a camera team from BBC News to make a half-hour special.

It was an intriguing time to be there because China was just beginning to open up to the rest of the world after the isolation imposed by Chairman Mao during the Cultural Revolution. Margaret Thatcher had recently paid an official visit – the first British prime minister to do so – and, even more significantly for young Chinese, Wham! had been the first British pop stars to be granted permission to perform there.

Things were definitely changing, but I had no idea what to expect – my impressions of China were still rooted in images of Mao's Red Guards, his Little Red Book and Communist Party mass rallies in Tiananmen Square. All that seemed to have vanished as we wandered round Chengdu amid a vast sea of people and found ourselves objects of great interest – large, incredibly well-organized groups of schoolchildren and swathes of men and women on bicycles were staring at us because they rarely saw Westerners in the flesh.

Billboards carried no advertising but instead urged parents to have only one child, and a huge statue of Mao still dominated the city's main square – an indication that the recent past would take a long time to erase. In fact, when I returned to Chengdu a couple of years ago, the statue was still there, but the city centre was completely Westernized.

Chengdu now produces half the world's microchips, has thousands of millionaires among its 16 million people and has the world's largest shopping mall, the New Century Global Centre, where I have lazed on the 160-yard-long artificial beach next to its fake Mediterranean village.

But how could we know back in the mid-1980s whether this new sense of quasi-freedom would last? Indeed, four years later the ruthless quelling of protestors in Tiananmen Square proved just how fragile it really was at that time. Jack (we never discovered his Chinese name as he always used the one given to him by his English teacher) was our young interpreter.

Since he was provided by the government, we were concerned that he could be a spy, so at first we were cautious with our words. But as the trip progressed and Jack became increasingly helpful in our battles with the (to us) unfathomable Chinese bureaucracy, we became good friends. And if his extra job was to spy on us, he had very little to report.

But he did tell us of his own family's experiences during the Cultural Revolution – a story which I didn't report at the time because I didn't want to put him in potential danger. Now it is safe to reveal that during the revolution Jack's family was forcibly separated for the simple reason that his mother and father were academics. They were each sent to different parts of the country and for eight years he didn't know whether his parents or his younger sister were alive.

When the revolution finally ended he was in his late teens

and he managed to make his way back to the family home in Shanghai. Apart from layers of dust, the house was just as it had been when they were taken away, even to the unwashed pots in the sink.

Jack lived there alone for many weeks, still not knowing what had happened to the others. Then suddenly, and without warning, his mother was at the door after being released from a detention camp. Both were amazed to see each other and, as you can imagine, were totally overcome with emotion.

Sometime later his father arrived, again without warning, and then finally his sister found her way home. The whole family was reunited, psychologically bruised but physically unharmed, unlike millions of others who were forced to leave their homes on the orders of Chairman Mao and his henchmen and never met again.

After showing us the sights of Chengdu, Jack and a team from the panda rescue squad drove us for fifty miles along bumpy roads and through remote villages (an indication of the poverty level was that rats were for sale as meat on market stalls) until we reached the Pitiao river, flushed with melted snow-water from the nearby misty mountains.

We were not far from the border with Tibet. Here it rains for 300 days a year and the steep slopes are clothed in bamboo forests. Or rather they were until two things happened – the bamboo famine and the ever-encroaching peasant farmers hacking down what was left of the bamboo on the lower slopes to grow crops.

On the banks of the Pitiao, in the Wolong national nature reserve, a rescue centre had been set up, and behind bars in its enclosure were five pandas that had been captured and were being brought back to strength. Their cages, open to the sky, were grim and basic but they could look up to their hills of home, from where every day their keepers brought them fresh supplies of bamboo. They were also fed with porridge – yes, bears eating porridge, but this was no fairy story. It was packed with nutrients and vitamins and they loved it so much they licked their bowls clean.

They had been caught in traps or sedated with tranquillizer guns and the hope was that if they regained full fitness they would be released back into the wild. The search went on to find any more that were starving up on the mountainsides and we joined a 'panda patrol' crossing a rickety rope bridge over the river and trekking high up the hillside towards the snow line looking for any sign of the elusive black-and-white bears. Their markings act as remarkably good camouflage in the dense bamboo jungles and if there were any we didn't spot them – sadly not even a bleep from a panda which had been fitted with a radio collar some months earlier. It had probably died beyond the range of the detector.

Bamboo must be the most ridiculous food any animal could choose to eat, especially one designed by nature to be a meat eater. For some unknown reason, millions of years ago, pandas with their carnivore digestive system switched to bamboo, which makes up 98 per cent of their diet and contains hardly any nourishment. They have to munch through about fourteen kilos of it a day just to keep going and they sleep the rest of the time.

Not only do they eat the wrong food, when it comes to reproduction they are pretty useless as well. Pandas are solitary creatures. There could be perhaps three of them on a mountain but they keep their distance – one of the Chinese vets described them to me as 'being alone together'. Females come on heat for thirty-six hours a year (yes, that's right, a year!) so by the time a male has eaten enough bamboo to gather the energy to find her it is often too late. If they do mate, naked offspring are born the size of a mouse, nine hundred times smaller than the mother. If she has twins she leaves one of them to die because she can't cope. Add to that the ever-increasing human 'invaders' in their territory and no wonder giant pandas are in such trouble.

I left Wolong wondering whether they would survive into the twenty-first century. But I underestimated the one great asset they do have, and that is their fantastic public relations appeal. With their endearing ways and cuddly image (don't be fooled, they can be really vicious), the world wants them to survive.

Emphasis has switched from rescue to captive breeding. Five years after I left Wolong the first baby was born there, and since then more than 300 have been reared successfully across south-west China. Most have been conceived through artificial insemination in specially built breeding centres, in particular the Chengdu Research Base for Giant Panda Breeding which is also, as you would expect, a huge tourist attraction. The Wolong centre was hit by the terrible earthquake in Sichuan in 2008 and has now been rebuilt. In the wild, numbers have increased to 1,800 in sixty-seven reserves and the species has been taken off the red list and is now classed as vulnerable.

In recent years I have taken groups of British travellers to panda breeding centres in Sichuan and, much to my surprise, was appointed UK Ambassador for Sichuan Pandas at a little ceremony in Chengdu.

My reward was to meet the latest batch of new arrivals – seventeen of them – at close quarters, something no tourist is allowed to do. But tourism is vital, because the visitors to the breeding centres – which are 'panda paradise' compared with the original cells at Wolong – help pay for the research work. Some wildlife experts argue that the money spent on saving pandas should instead go to preserving other species which stand a better long-term chance of survival. But what will future generations think of us if we allow this much-loved symbol of conservation to disappear on our watch?

19. Editor of *Newsround*

On Wednesday, 19 November 1986, the phone on my office desk rang just before that day's *Newsround* was about to be broadcast. It was a reporter from the *Sun* newspaper asking for my comments on a tip they had that I was getting the sack because of my age. It came as a bolt from the blue but all I said was 'It's news to me' before heading for the studio, a rather worried man.

Some changes had been taking place in children's television after Edward Barnes retired as the boss and had been succeeded by Anna Home from the drama side of the department, but I hadn't heard of any threats to my position. When I got back to the office there were reassuring calls from both the BBC press office and senior staff, and late that night I even got a call at home from the BBC One controller, Michael Grade, who was in New York, telling me the story was nonsense and my job was safe. He'd always called me his 'five o'clock friend' because, he said, with me in the chair he could be guaranteed a good audience figure at teatime.

Even so, the next day's banner headline on Page Three of the *Sun* was: 'John Craven Out – He's Too Old at Forty-Five'. Under an 'exclusive' tag the story read:

> Children's favourite John Craven is to lose his job as presenter of TV's daily *Newsround* programme – because BBC bosses think he is too old at forty-five.

> Youthful looking John has been anchorman of the popular show since it started in 1972. Viewing figures often reach 9 million and the bulletin is considered one of the Beeb's big successes. But the new head of BBC Children's programmes Anna Home wants a younger presenter to encourage more kids to tune in.

The story was not followed up by any other paper, thus confirming that there was nothing in it. What was really happening was that I was in negotiations with Anna to take over from Eric Rowan as the editor of *Newsround*. Far from being sacked I was taking on new responsibilities.

Because of the extra duties and demands of the new job I wasn't able to present the show every day so two of our talented reporters, Roger Finn and Helen Rollason, would step in for me on days when I was bogged down with BBC bureaucracy.

Helen had joined us to do sports reporting but quickly took on a much wider brief and was hugely popular. When she left the show in the early nineties she became the first woman to present *Grandstand*. Helen sadly died of cancer in 1999 and is still very much missed.

Roger was born and brought up in Hong Kong but went to boarding school in the UK and on some Sunday lunchtimes the headmaster gave him special permission to listen to the radio – so that he could hear his mother! June Armstrong-Wright was a star broadcaster in Hong Kong and one of the voices on the BBC Light Programme's hit show *Two Way Family Favourites,* which linked British forces overseas with their families back in the UK through their favourite records. While his mum was reading out requests six thousand miles away, young Roger was being reminded of home.

Spin forward twenty years and he was on broadcasting duty when the American space shuttle *Challenger* blew up on take-off just a few minutes before *Newsround* went on air. Adult news bulletins had become bored with routine shuttle missions but we had been following the build-up to this one very closely because a schoolteacher called Christa McAuliffe was on board.

Christa had been chosen from 11,000 applicants to be NASA's teacher-in-space and was taking with her projects from children all over the world, so this mission was a natural for *Newsround*. I had not been in the office that afternoon and was in fact making my way back there from the Stock Exchange, where I'd been discussing a possible *Newsround Extra* about the way the City works.

At five o'clock I paused in the street outside a television dealers just to make sure the programme was safely on air and to my shock and disbelief there, on dozens of screens, was the *Newsround* opening sequence with *Challenger* exploding like a firework in the deep blue sky. I quickly nipped into the shop but the sets had the sound switched off so I couldn't get any detail.

But there was Roger giving out what was obviously disastrous news and handing over to James Wilkinson, the BBC's science correspondent, for more details. From the pictures all around me it was impossible to tell whether the shuttle was still in one piece, but the situation looked bad.

'I was in the studio going through the day's scripts when I heard in my earpiece that *Challenger* had blown up' says Roger. 'Our producer had requested a recording of the launch because of the Christa McAuliffe connection so we had the pictures but very little information. I was really nervous as it was only about my sixth time in the chair and so hugely aware that millions of children would be watching. Christa was an inspirational figure to them and I had to get the tone right.'

He did it well – it's not the easiest thing to tell the nation's children that a teacher they had heard all about on their news programme had been in a terrible accident right at the start of what should have been her space adventure.

Amid all the controlled chaos that hits a programme at such times he remembered the BBC One newsreader Julia Somerville coming into the studio and a classic discussion taking place about whether News or *Newsround* should break the story. *Newsround* won, consolidating the important principle that we had fought for over the years, which was that if a major news event happened during children's time on television we were the proper outlet for it.

Since the early days *Newsround* had been interested in space for the simple reason that our viewers were keen on the subject and they had been kept very well informed by two 'grandfather' figures who had vast knowledge of the subject – first, and briefly, Patrick Moore and then, for much longer, Reginald Turnill. Reg had been the aviation and aerospace correspondent for BBC News until he retired and we gave him a second career as *Newsround*'s space editor. He loved it, and would catch a train from his home on the Kent coast to bring our audience the latest, often exclusive, news about astronauts and their missions.

He was extraordinarily well informed and knew everyone of importance at NASA on first name terms – Buzz Aldrin, the second man to step onto the moon, was a great personal friend. Reg formed a fantastic working relationship – and friendship – with *Newsround*'s youngest producer, Lewis Bronze, who was twenty-three when he got the job, and together they produced some great scoops. Lewis – who later became editor of *Blue Peter* – paid a very moving tribute to a man he really admired at Reg's funeral some years ago.

I liked and respected Reg for his professionalism, his often unexpected humour and his way with words. He was in a restaurant near Cape Canaveral and the waiter brought him a lobster with one large pincer missing. When asked where it was, the waiter said sarcastically: 'Lobsters get into fights and sometimes they lose.' 'In that case,' said Reg, 'please take this back and bring me the winner.'

When he finally retired, he gave a party at Broadcasting House for colleagues over the years, including his more recent friends on *Newsround*. In his speech Reg made a few disparaging, but very funny, remarks about the bosses he had worked for and ended by saying: 'People tell me the BBC isn't what it used to be. I have worked for the BBC for fifty years and let me tell you – the BBC never was what it used to be.' No rose-tinted glasses for a man who looked into space!

One teatime in 1986 a new piece of kit appeared on my studio desk alongside the trimphone. 'This,' I explained to my audience, 'is a computer.' It was, in fact, a terminal that was connected to a mainframe computer that was beginning to revolutionize broadcast news. It could do many things: provide the latest news stories, be used to type scripts and the autocue – in fact, it did away with all the journalistic paperwork. It had recently been installed at Television

Centre and, like other new techniques, was being tested out on *Newsround*.

Which is why John Humphrys and co. were standing in the corner of my studio to see how the system worked on air. Until now, the autocue had been typed out on a long strip of paper which was fed through a machine and newsreaders could make last-minute changes, either by hand or by dictating to the operator. Now, everything was done through the mainframe computer and it worked a treat.

The system was designed to cope with late changes and, after being test-bedded on *Newsround*, it went into action across the entire BBC news and current affairs division.

Newsround was also the first bulletin to use videotape rather than film in its cameras. Back in the late 1970s Eric Rowan and I were in New York for some *Newsround Extra*s and one was about soccer, which was already big in American schools but was still struggling to make an impact professionally.

I got to meet Pelé, one of the greatest footballers ever, who was promoting the sport, and when we filmed a major game featuring the New York Cosmos, the huge screen, which I'd never seen before at a match, flashed up the message, 'The Cosmos welcome *Newsround*' followed by a massive pair of electronic hands applauding. Not sure what the fans made of that.

During that trip we were invited to look round the CBS News headquarters and shown what they called their 'black museum' of outdated news equipment. Among the exhibits was, to my surprise, a film camera. 'We only use video now – film is a thing of the past,' was the explanation, making us feel a bit like news dinosaurs. We also sat in the gallery for that night's transmission of the national bulletin fronted by the legendary newscaster Walter Cronkite, known as the most trusted man in America. He broke the news of President Kennedy's assassination and President Nixon's resignation, and when he reported from Vietnam that he thought the war was unwinnable it was said to have changed public opinion in the States.

With a minute to go, Walter was at his studio desk in his shirtsleeves, finely polishing his scripts. Then 'Thirty seconds, Walter, please get your jacket on,' shouted the floor manager. The great man took no notice and continued writing. 'Ten seconds, Walter,' called the floor manager desperately.

Slowly, as the last seconds counted down, on went the jacket and he adjusted his tie. By the final second he was ready for his cue. 'Good evening, America. This is Walter Cronkite with the CBS News.' That's how to do it. Totally cool. 'He does this to us regularly,' said our guide, 'but not for much longer. He's retiring soon and there's a plan afoot that on his last night he'll find his jacket sleeves have been sewn up.' I never did find out if that actually happened but one thing is for sure – he will have ended with his famous catchphrase, 'That's the way it is.'

Memories of Walter and his CBS pals came flooding back a couple of years later when we heard from BBC News bosses that they were in talks with the Japanese electronics firm Ikegami about video cameras and would *Newsround* like to test one out.

We used it for several weeks, with the big boys in News keeping close watch, and found it much quicker and easier to use than film. Shortly afterwards it was given the go-ahead on the main bulletins and top cameraman Bernard Hesketh took an Ikegami with him – maybe it was ours – when he was dispatched at short notice to cover the Falklands conflict.

It was the first time news had used a video camera in a war zone and Bernard won a BAFTA for his work. And so a new acronym came into the world of news: ENG, standing for electronic news gathering. The new age of newscasting in Britain had begun and every cameraman would soon be shooting on video.

Another pioneering innovation we tried out was a kind of light pencil which I could write with on a monitor screen, but sadly it turned out to be rather unreliable. When the mountaineer Sir Chris Bonington was on one of his Everest expeditions I used the light pencil to show viewers exactly where he was on a map of the mountain. Unfortunately it slipped on the screen and the 'X' that pinpointed Sir Chris showed him hovering a couple of thousand feet above the summit. Back to the drawing board.

After I became editor I had a regular weekly meeting with Anna Home about the content and staffing of the programme, and during the first World Aids Week Anna said she thought *Newsround* should tackle the subject – a brave, potentially controversial idea. But research suggested that children were hearing all kinds of misleading playground information about Aids – that you might get it if your dad kissed you, or from a toilet seat and other such nonsense.

So we set about putting things right without ever mentioning sex, which was quite an achievement. We said people who had to have lots of blood transfusions might be at risk but that problem was being dealt with and that adults who were in potential danger knew the precautions they had to take. And you certainly couldn't get Aids from kisses from dad. Basically, we were putting rumours to rest and telling our audience that Aids was a new illness which they personally needn't worry too much about.

As we were spending much of the programme talking about Aids I phoned Phillip Scofield, then the fresh-faced presentation star of BBC children's television, in his famous 'broom cupboard', and said he had best not be too jolly as he announced *Newsround* that night. 'Fine,' said Phillip. 'I'll make sure Gordon the Gopher isn't with me – it'll be just a straightforward introduction.' And the consequence of that was more complaints to the duty office from viewers about Gordon, the glove puppet who was almost as big a star as Phillip, failing to appear than there were about *Newsround* reporting on Aids.

After a couple of years editing the programme but also having to attend endless staff meetings and sit on appointments boards I had come to the conclusion that I was not cut out to be a bureaucrat. My real love was contained in the second word of the BBC's title not the third – I was a broadcaster not a corporation man. Also, the presenters on children's television were getting noticeably younger and my dark hair was showing signs of grey.

One day in late 1988 I looked from my office into the *Newsround* newsroom and the truth hit me that everyone else in the team used to watch the programme when they got home from school. I'd been doing the show for that long. It was time to move on before it was too late for me to escape from that pigeonhole which had trapped so many children's presenters. It was time to grow up and try to get an on-screen job in mainstream television after a gap of seventeen years.

I told Anna that I would not be renewing my contract and

she agreed that I could break the news about leaving while I was a guest on the latest Saturday morning show, *Going Live*. Phillip Scofield and his co-host Sarah Greene had not been warned in advance so it came as a shock to them and to the viewers.

The Sunday papers reported my departure and the next day BBC Birmingham phoned and offered me some work, which was generous of them because I hadn't even started looking for a new job. All I had decided was that it was time for a change, and if nothing materialized in broadcasting I could always go back to print journalism. Little did I know that the Birmingham call would lead to an even longer stint than *Newsround* when I became presenter of a new rural-affairs programme that was based there called *Countryfile*.

Throughout this uncertain time my wife Marilyn was, as always, like a rock – supportive and encouraging and sure that everything would turn out for the best. On my very last *Newsround*, I ended by saying: 'And finally, this is the last time I'll be saying "and finally",' and thanked everyone for watching me over the past seventeen years.

Those years had been an incredibly satisfying time professionally and I had worked with some lovely and talented people. It was also a very happy time – I can't remember even one serious bust-up among the team. Though people had come and gone, everyone had enjoyed themselves on the show and learned a lot about communicating in simple but effective terms. We were friends as well as colleagues.

What we created had become something of an institution and it is still going strong nearly half a century after that six-week experiment. It is very different from the programme I knew, serving a very different generation of viewers in a very different world. Now it is shown several times a day, not just once, on the CBBC channel rather than on BBC One, and though the real-time viewing figures are a fraction of what they used to be, today's children also watch it on their iPads, on catch-up and on the impressive *Newsround* webpage. And the list of presenters who followed me and went on to forge successful careers in prime-time television is truly impressive. They include Krishnan Guru-Murthy (*Channel 4 News*), Julie Etchingham (*ITV News at Ten*), Juliet Morris, Helen Skelton, Matthew Price and Adam Fleming (BBC correspondents), Jake Humphrey, and Ore Oduba, who won *Strictly Come Dancing* in 2016.

Its mission has not changed over the years. It informs and explains in an eye-catching, non-condescending, inclusive way and with so much fake news around and so much disturbing material on social media, we as a nation should be grateful that our children still have *Newsround* to turn to.

When I left there was a great farewell party thrown by the team, and Paul Fox, the big boss of BBC Television, also hosted a dinner party in my honour in the plush sixth-floor suite at Television Centre. At one point it felt almost like a retirement dinner, so I had to remind everyone around the table: 'Hang on. I'm not finished yet!' Or so I hoped.

20. *Countryfile*

One morning in the mid-1990s I strolled into the canteen at the BBC's Pebble Mill studios in Birmingham and came across a large group of children in their first years at a local primary school. They were on a tour of the building and filling the place with excited chatter. I went up to say 'hello' and the looks on their faces said it all – 'Who's he?' Not one of them recognized me, which was rather a shock because during my time on *Newsround* just about every child in the country knew who I was.

Anyone who is a regular face on children's television stays in the minds of viewers during their developing years and way beyond! In the nicest way there is no escaping from them. Nowadays for me the surprise comes when strangers who I think look rather older than I do come up and say: 'Thank you for being part of my childhood.'

But back then I had been away from children's television for five or six years and hardly any youngsters watched my new programme, *Countryfile*. It wasn't intended for them and was transmitted on Sunday morning, when children were busy doing other things. For many years after I joined the programme if a child did ask for my autograph it was in a rather puzzled way and nearly always because they had been encouraged to do so, usually by their mother or their granny who wanted it for themselves.

Not until *Countryfile* moved to peak time on Sunday

evenings in 2009 did children start watching with their families in large numbers. Now if they ask for an autograph or a selfie it'll be because they want one and so, in terms of recognition, my career has gone full circle – and, like *Newsround*, *Countryfile* has become something of a national institution.

Within days of leaving *Newsround* in 1989 my base had moved from Television Centre to Pebble Mill, and for a long time I really missed the old 'dream factory'. Every working day for seventeen years I had walked through the gates of that iconic (when I started writing this memoir I promised myself I would ban that hackneyed word but in this context it really is the only one) building and thank my lucky stars. In fact a high point of my career was when I was issued with a permanent pass for the very-limited-space car park at the front of Television Centre. I felt I had really made it.

I could not have hoped for a better workplace and it had given me some wonderful opportunities and memories. If I had been working late I would often wander along the second-floor circular corridor and peek into the observation galleries and see what was going on in the main studios. Through the big glass windows you could look down on all the action.

In one studio there might be a major star-packed drama and you could catch the informal moments between takes. Another studio would be getting ready for a recording tomorrow with floor painters, set builders, electricians and scene shifters hard at it in this fantasy land.

Walk into the next gallery and a warm-up comedian could be persuading the audience to laugh, cheer and applaud at the silliest of quiz shows. And so it went on, something different in each studio but each contributing to the output of the best television network in the world. How sad it was when the BBC, in its wisdom, decided to move out. Now the building is apartments and hotel rooms, though three of the studios have been brought back to life.

What a contrast there was to the perceived glamour of Television Centre when I joined *Countryfile*. No make-up artists to powder my nose, no floor managers to keep things under control, no big lights shining down and directors calling the shots in my earpiece amid the exhilaratingly tense atmosphere of a live studio.

Instead I found myself miles away from anywhere, traversing the hills and dales of rural Britain, with the wind snatching my words away, the rain making life difficult and the sunshine making me realize I had probably the best job in the world.

My career so far had been in news. I had been a deadline junkie, metropolitan-based and chasing headlines. Now I was swapping headlines for hedgerows as I wandered along country roads bringing into focus many aspects of life 'out in the sticks'. As if to underline the change to my working life, *Countryfile* was recorded on location not transmitted live.

Yet I was still a 'newshound' because from the very start *Countryfile* demanded a high standard of journalism. And another thing did not change – I could still wear casual clothing. For nearly fifty years now I haven't had to put on a suit to work; for *Newsround* it was always jumpers and shirts and for *Countryfile* I switched to jumpers and anoraks because it was, and still is, the brand leader of what came to be called 'welly telly'.

Many other shows on all channels now urge people to get up from their comfy chairs, don sturdy clothing and enjoy all that our countryside has to offer. But *Countryfile* led the way – and for three decades it has given me the greatest joy to be out there in all weathers encouraging viewers to follow in our tracks.

The programme had been on the air for exactly a year when I became part of the team. Journalistically, of course, my style had to alter – I was now broadcasting to grown-up people like farmers and other country folk as well as townies who were interested in the countryside.

But I was determined to stick to my old mantra, 'Keep it simple without being simplistic'. Over the years *Countryfile* has shown me every corner of the British Isles, sometimes more than once, and when I'm on my travels it's often in the opposite direction to all the traffic – away from the towns and into the wide open spaces.

Often my destinations are so remote you don't see another vehicle or another person for mile upon mile – on these crowded islands, the other sixty-six million are somewhere else. Perhaps the greatest privilege is to be given access, with our cameras, to places where few are allowed to tread and to show these secret spots and their natural treasures to our viewers.

To my initial surprise I discovered that a large section of the farming community in those early days actively resented *Countryfile*. For many years they had had their own show, called simply *Farming*, which examined their industry in great detail. Other, non-farming viewers who happened to have their television sets switched on at that time of day could 'eavesdrop'. The BBC not only cancelled this much-respected programme but, to add insult to injury as farmers saw it, put the upstart replacement *Countryfile* in the same time slot at Sunday lunchtime.

Like its predecessor, it was broadcast almost every week of the year, and though farming was at its heart, the new show took a much wider view of the rural scene, examining the social, environmental and economic issues facing country folk and even questioning some farming practices from an outsider's point of view.

Michael Fitzgerald was the first series producer and he and I went back a long way – we had shared digs in Beach Avenue, Whitley Bay when I moved to the north-east: he worked in the BBC newsroom in Newcastle and it was to Michael that I first confided that I would like to be a broadcaster. Now he was confidently leading *Countryfile* in a direction that some of its core audience were far from happy about.

When asked at the time about the transformation Michael told *Radio Times*:

> We had to ask ourselves if the original programme still made sense, bearing in mind the stresses and strains the countryside is currently facing. Villages are now very desirable places to live in but there's a shortage of low-cost housing.
>
> There's also a huge debate on access to the countryside and so-called increased leisure. These issues all affect the farmer, but we hope to broaden public awareness. I like to

think that we'll revel in the joys of country living but we'll also think of the implications. We won't, by nature, be campaigning but we will give access to people who have something to say about certain environmental issues.

Those were the original ground rules and they have stayed much the same to this day. Michael's reporting team was made up of naturalists, environmentalists and journalists – Chris Baines, Roger Tabor, Anne Brown, Caroline Hall and Ian Breach – and the first show on 24 July 1988 included features on canoeists who were demanding more access to rivers, the shortage of affordable rural homes and cats threatening small wildlife.

To mark the twentieth anniversary of the show we re-examined all the issues raised in programme number one and found that not much had really changed. It was also on that anniversary show that I had the strange experience of seeing myself standing behind me. It wasn't an illusion – it was an impressionist, the brilliant Jon Culshaw, star of *Dead Ringers* and many other shows.

I was in a field in Swaledale in the Yorkshire Dales, just going over my script for the next introduction, when I heard my voice repeating the words, and when I looked over my shoulder there was a mirror image of me. It took a couple of seconds to realize that it was Jon doing a very accurate version of me, using some jerky mannerisms I never realized I had. We had a good laugh and later I discovered the lengths the team had gone to to make sure the surprise worked.

Jon's only props were a greyish wig and versions of the blue jumper and red anorak that I often wore on the programme. Andrea Buffery, the director, had telephoned Marilyn before

I left home saying: 'Can you make sure John wears his blue jumper and red anorak?' and explained why.

When Marilyn, keeping the secret, suggested the outfit to me I didn't like the idea because I had worn it on the previous week's show.

'Well, take it anyway, just in case anything happens to the ones you will be wearing,' said Marilyn, and I reluctantly agreed. Once in Swaledale, Andrea found some excuse to get me to put on the red and blue, and the stunt worked a treat.

Since then, Jon has impersonated me many times on his radio and television shows and usually he is kind. But I admit I wasn't too sure about his sketch involving me and Bill Oddie in a tent on Brokeback Mountain! One day I was at Oxford Circus Tube station and felt a hand on my shoulder. It was Jon and he said: 'The last time I saw you, I *was* you.'

Though he's very good at looking and sounding like me, there was a time when some people thought I bore more than a passing resemblance to Bill Clinton. The *Evening Standard*'s distinguished critic Milton Shulman wrote in 1992:

> 'While aware that it is difficult to get away from the American presidential election on TV, I was taken aback last Sunday at seeing the Democratic candidate Bill Clinton in the Highlands of Scotland talking about crofters. Closer examination, however, revealed that the man in the windproof jacket and blue sweater was not Clinton but John Craven, the frontman for the BBC's *Countryfile* programme. If Clinton becomes president and ever needs a double for some secret mission then John Craven would be ideal for the role. On TV they look like identical twins.'

I never got the call! But Mr Shulman might have liked to know that when Marilyn and I were on holiday in Egypt, hawkers chased us through a souk shouting; 'Mr Clinton, Mr Clinton – please buy, please buy!' Hardly a secret mission, but it proved the likeness had international recognition.

Crucially, back on that very first *Countryfile* in 1988, the weather forecast for the week ahead was retained because it was essential information for farmers and anyone else out in the countryside. Bill Giles compiled the first forecast and thirty years on it is still a vital element of the show.

Just about every BBC weather presenter has had his or her turn and I know they take particular pride and care in preparing it because millions of country folk rely on their predictions. For that matter, so do millions of townies. The forecast is the only live element of the show and viewers often tell me they think it's one of the best bits.

On my first day on *Countryfile*, how prepared was I to become the lead presenter of a current affairs programme set amid the green acres? I was across many of the conservation and environmental issues the programme tackled because of my *Newsround* background but I knew very little about the core ingredient, agriculture.

I had to learn quickly because there were farmers around the country who, in no uncertain terms, had no time for people like me on their land asking what they thought were pointless, ill-informed questions – and they wanted the old programme back.

To some extent I could sympathize with their attitude but no other industry had its own designated television programme, with the possible exception of *Top Gear* and the motor industry, and *Countryfile* was encompassing agriculture as part of its wider rural brief.

Farming can be a lonely and financially uncertain job, and to those cynics who say they've never seen a poor farmer I can honestly reply: 'I've seen lots who barely scrape a living and some who have had to give in and give up.' When problems mount in this unpredictable industry they can weigh heavily on the mind if you're working all day in the fields with no one to talk them over with.

Many farmers, especially the older ones, tend to be secretive souls who wouldn't dream of discussing their worries with their family, friends or neighbouring farmers let alone with me. This reluctance to confide, to seek help when it is most needed, is perhaps why the industry has one of the highest suicide rates in the UK.

Over the years I have made a point of asking farmers whether their sons or daughters will be succeeding them, and all too often the answer is no. They don't want the long hours and uncertainty and it begs the question, 'Who is going to be there to farm the land in the future?' Many rural colleges removed the word agriculture from their titles, such was the lack of interest.

Slowly things are beginning to change for the better. Some colleges have reintroduced their agricultural courses and students, especially those from non-farming backgrounds, are realizing there could be a good future in this industry as world demand increases for food and Far Eastern markets continue to develop a taste for Western crops.

It took some time, but eventually many farmers came to the conclusion that *Countryfile* was a decent platform on which to present their concerns, and even their personal feelings, to the public at large. We certainly weren't a pushover and would ask tough questions when necessary. But as farming began to open its gates to us, both metaphorically and physically, our viewers realized that by and large there is not much wrong with British agriculture.

Welfare standards are among the highest in the world and a great number of farmers do genuinely regard themselves as custodians of the countryside – things that matter a great deal to an increasing number of our viewers who want to know how their food is produced, where it comes from and how that process affects the balance of nature.

But why has a straightforward, un-gimmicky and always low-cost programme about our countryside gained such a place of affection in the nation's heart? For me it's always a lovely, satisfying moment when complete strangers approach and say they love *Countryfile* and try never to miss it. That devotion has been engendered, I believe, because it is good, old-style family viewing with interesting content, no swearing and no sex, unless it involves animals at a discreet distance.

From the earliest days it had a respectably sized audience, certainly for Sunday morning, of around two million, even though it was not shown in Scotland and often not in Wales, and that figure grew when we stretched to sixty minutes a week and went nationwide in 2003. Six years later we were moved to prime time, the audience trebled and dear old *Countryfile* found itself beating shows like *The X Factor* and *Poldark* in the crucial Sunday night ratings. I have come to the conclusion that the standout star of the show is the British countryside itself.

The presenters usually have beautiful views over our shoulders and the show is packed with shots which make me personally proud to live amid what I think are the loveliest landscapes in the world.

Recently a writer in *The New York Times* (of all newspapers, but *Countryfile* is now shown in the United States on the streaming service Britbox) commented on the viewing figures and added: 'Some achievement in a land where the battle for TV ratings produced a dating show in which participants appear naked. If there is anything raunchy about *Countryfile* it is perhaps best described as pastoral porn: shots of bare, undulating hills and deep, grassy valleys'. I smiled over that description of 'pastoral porn'. It was the one and only time that I have seen the words 'porn' and '*Countryfile*' in the same sentence!

I've long held the belief that we British take a proprietorial attitude towards the countryside. We don't own it – at least, most of us don't – but we behave as though we do and we want to protect it. Wildlife and conservation groups have millions more members than political parties.

Many of us have a rural gene that goes back many generations – after all, until the Industrial Revolution most of our ancestors would have been country folk. Our audience is split down the middle between urban and rural viewers and our challenge is to appeal to both sides without appearing elitist to the former and condescending to the latter.

The modern countryside has two separate purposes – on the one hand it is an expansive, efficient and sometimes noisy workplace and on the other it offers a landscape of great, often majestic beauty which, as a place to live or just to spend time in, satisfies the soul. And though those purposes are separate they intertwine – 70 per cent of the land we love has been created by farmers while tourism provides a huge boost to the rural economy and gives farmers the chance to diversify. I doubt if either could survive without the other and it is this homogeneous, holistic image of rural Britain that is portrayed on *Countryfile*.

It has never been a campaigning programme – presenters are not allowed to bang their own particular drums, although, for example, a vegetarian would not be expected to report from an abattoir. We have always believed in presenting a balanced view on any topic and the evidence we produce on screen helps our audience to make their own personal decisions as sensible adults.

My first report for *Countryfile* in July 1989 was an investigation into the future of organic farming. 'Does it have one?' I asked, a question that is still being asked today and a subject I have returned to many times. Supporters believe it certainly does and that it could be even more viable if the public was prepared to pay a little more for their food and show a little more concern for the way it is produced.

Certainly, the market for organic food is larger now than it was in the late eighties. Supermarkets are often portrayed as the modern-day wicked barons of rural Britain, dictating how food is produced and in what quantity and holding many farmers virtually to ransom if they want to keep their contracts.

Some of those criticisms can be justified, but these days I do see a wide variety of organic produce on their shelves – something that wasn't there when I compiled that original report. But the problem for niche markets like organics will always be that many customers who want to be as ethical as possible tend to park their principles when money is short.

After that first report, which took up the whole show, I decided I really enjoyed my new role and that *Countryfile* was for me.

It quickly became obvious that the programme needed a

four-wheel-drive vehicle to reach some of our locations. In 1989 Land Rover launched its Discovery model, which was just what we needed, so one of the first was acquired for the show and put to good use almost immediately.

I drove it from Birmingham to the Cairngorms during a snowstorm in the middle of winter. Conditions were so bad and the snow so deep that the A9 road through the Highlands was about to be closed. The Discovery and I just made it through but it was a hairy drive and I was mighty pleased to reach the safety of the hotel car park in Aviemore.

Next morning, as I was brushing away the overnight snow from the *Countryfile* emblem painted across its large bonnet, a very grumpy guest watched me and then launched into a totally unexpected tirade. 'Why is the BBC buying foreign rubbish?' he fumed. 'Is this what you're spending my licence fee on? Why can't you do the decent thing and buy British?' Obviously he thought the Discovery was from the Far East and though I tried to explain that it was the very latest model from a British production line he wouldn't listen. Instead he stormed off – in his Citroën!

The emblem, involving a large butterfly, was on the bonnet and along the sides of the vehicle because the powers-that-be decided *Countryfile* needed a mobile 'set' to emphasize the programme's identity. So for several years I leant on the bonnet or looked out from the driving seat, with emblems all around, to do my introductions. Eventually we (and probably the viewers as well) got bored with that and the emblems disappeared. But we have used a Discovery in its various forms to get us to the wildest places until recently when, to emphasize our 'green credentials' we switched to a hybrid four-by-four. It is foreign, but not rubbish!

One Discovery was involved in an off-screen drama when it freewheeled down a steep Welsh hillside. Director Mick Murphy, who was in the driving seat, managed to stop it just before it plunged into a river. 'I saw my whole life flash in front of me – badly edited!' he later joked.

Mick is a great impersonator (he takes me off better than anyone except Jon Culshaw) and I hate to think of the number of times he has pretended to be me. He once phoned Marilyn and chatted away with her in my voice for a couple of minutes. Then she said: 'You can stop now, Mick – I know it's you and not John.'

'How do you know?' asked Mick.

'Because he never calls me darling on the phone,' replied Mrs C.

Mick was brilliant at impersonating noises as well as voices. He and I were making a *Countryfile* report about proposed boundary changes and as we wanted to trace the history of English counties we were given special permission to film the Domesday Book at the National Archives. We were led down corridors to a secure room and there it was on a table in front of us, the great book itself. I was handed white gloves and while a security guard looked on and an archivist gave advice I was allowed to open it and turn a couple of its 913 parchment pages.

What a privilege we had been given and I spoke to the camera about the significance still today of this survey of England's wealth ordered by William the Conqueror. After the take, Mick said: 'That was great, John, but just to be safe can you do the whole thing again? I really liked it when you turned the pages.'

So we did a second take and as I lifted a page he made the sound of it being ripped. It was so realistic, and I had been concentrating so hard, that fear shot through my body. I had inadvertently damaged what is probably this country's most famous book! Only when everyone else in the room burst into laughter did I realize I'd been tricked and the Domesday Book was still in its original, 1086 condition.

For well over a thousand editions I have traversed our green acres by car, four-wheel drive, tractor and aeroplane. I hate to think how many gallons of fuel I have used and how many emissions I have created – it's an environmentalist's nightmare, but how else could I get to truly remote places? I use the train whenever possible, but long gone are the days of the rural branch line. Heritage railways have taken over some of those routes, and I have filmed on several of them, but rarely do they offer a round-the-year regular service.

There's no doubt that some of the finest views of our countryside are from the air, and I was amazed to discover, when I made a film about how rural Britain looked at the beginning of the Second World War, that the best contemporary aerial photographs of East Anglia were taken by the Luftwaffe, the German air force.

Their reconnaissance planes flew at 30,000 feet with pilots equipped with oxygen masks and powerful cameras. While the Battle of Britain was going on far below them, they were searching for the sites of new RAF airfields and military bases.

We managed to get hold of some of those images and they were pin-sharp and quite stunning – a time capsule of how things used to be. We flew over the same areas in a small plane and I expected the landscape to look very different, but it didn't.

Some villages had grown into towns, but even in 1940 East Anglia had many large fields – a legacy from a century earlier when the area pioneered the use of massive steam ploughs which needed lots of space. One notable absence in the Luftwaffe photos was Stansted Airport – that didn't start life, as a US Air Force base, until two years later.

Much has happened in the countryside since I started doing the rural rounds rather than the news rounds. Let's look at one of the positive aspects first – food in country pubs. Back in the early 1990s you were lucky if you could get a sandwich at lunch-time and dinner was often out of the question. Pubs were there for beer and conversation and strangers could expect fierce stares from locals should they step into their domain.

But slowly things improved because they had to – too many pubs were, and still are, going out of business at an alarming rate – more than 10,000 of them across the country so far this century. Nowadays you can get really good meals in quite ordinary looking pubs, not just those trendy gastro establishments with an eye on a Michelin star.

If you're lucky, you can eat at any time during opening hours – something that's vital for film crews working unpredictable shifts – and the food usually comes within our limited meals allowance. It's not just ham, egg and chips or sausage, egg and chips on offer. There is a huge choice, and though some items on the menu may be home-warmed, after arriving in large vans, rather than home-cooked, in my experience it is nearly always good quality.

But I was once sitting at a table just by the kitchen door and heard one server say to another about a customer: 'He actually ate it!' The connotations of that just don't bear thinking about. During my early years on *Countryfile* one of the few places you could get a hot meal at any time of the day was a Little Chef and I developed a soft spot for these roadside cafés and especially their Jubilee pancakes filled with cherries. With more than 400 Little Chefs to choose from they became such an integral part of our travels that I got to know the range of food by heart. In fact, I used to say that if ever I was invited on to *Celebrity Mastermind* my chosen subject would be Little Chef menus. But I changed my mind when Heston Blumenthal was asked to revise them, because that would make the questions far too difficult. Now, sadly, Little Chefs and their memorable pancakes are no more.

I have spent hundreds of convivial evenings in country pubs with *Countryfile* teams after a long day's work and it's hard to find better places in which to unwind. It's not just pub food that has got better – so has accommodation. Rooms now are usually en suite, with a smart selection of toiletries and soft towels. Not like some places I stayed in at first when you had springs sticking out of the mattress, long walks down freezing corridors to reach the bathroom and it was by no means certain the water would be hot or the loo would flush properly.

From time to time I stay in farmhouse B&Bs, mainly because the money the farmer's wife makes from taking in guests can play a large part in keeping a small farm going – something which became patently obvious when the disastrous outbreak of foot-and-mouth disease in 2001 put the countryside more or less out of bounds and visitors stayed away for many months.

You always get a cheery welcome when you're checking into a farmhouse – those early jibes about *Countryfile* have long vanished and, after all, you are putting an extra few quid into the rural economy. A colleague tells how he arrived unexpectedly at one farmhouse late at night. His shoot had overrun and he didn't want to drive home tired so when he spotted a vacancy sign he knocked on the door.

The farmer's wife welcomed him and said, 'I've got two

en-suite rooms. One has a shower and the other has a bath so it's a bit more expensive.' When he asked what the difference was she gave him a look that implied he must be an idiot and said: 'Well, you have to stand up in a shower.'

Two electronic innovations were to make our lives much easier as we criss-crossed the countryside – mobile phones and, later, sat navs. In my first years on the show it was nigh impossible to get in touch with the office because phone boxes were so few and far between, especially ones that worked. Sometimes we would arrive at a location only to discover it had been changed but no one had been able to contact us.

A friend had splashed out on a mobile and I thought he was just showing off. 'Not at all,' he said. 'Get one yourself and the first time it rings in the middle of nowhere and the call turns out to be really important you will realize just how vital this little gadget is to you.' At the time it certainly wasn't little but he was right.

Until sat navs came on the scene I was forever having to stop the car and check the map as I attempted to make my way to remote locations. Sat navs changed everything and I will not hear a word said against them, even though one did try to guide me into the sea near Weymouth. People sometimes ask me what route I took to get to their isolated place and when I tell them they look surprised and say only locals know that way. 'And the sat nav' I reply.

My arrival at *Countryfile* happened to coincide, unfortunately, with the start of a string of deadly diseases which hit the nation's farm animals. The list seemed endless: BSE, anthrax, salmonella, listeria, bovine TB – and everyone started to wonder if anything worse could possibly happen. It did, of course, and I still remember the sight and smell of those awful burning pyres as foot-and-mouth scythed through the landscape in 2001.

Fifteen years before that disaster many cattle farms were hit by BSE – bovine spongiform encephalopathy – which was tagged 'mad cow disease' because of the distressing way infected animals staggered uncontrollably as it began destroying their brains. In France (which did not recognize the scale of its own outbreak until much later) it was known as 'le staggers'.

When a case was discovered in the UK the animal was slaughtered. Altogether 180,000 cases were reported and another 4.4 million cattle were destroyed as a precaution, at a cost to the economy of around £900 million. In some places on the Continent, so the rumours went, if an animal staggered it was secretly shot and buried in a field to avoid notification. I filmed for *Countryfile* in the European country which, after the UK, had the highest number of animals officially destroyed– and that was little Switzerland. Like us, the Swiss stick to the rules.

To begin with, farmers were baffled – no one knew what caused the disease and how widespread it would become. Government scientists were called in and eventually the blame was pinned on the processed meat-and-bonemeal being fed as protein to herds. Some of it was being made, it was later discovered, from the remains of infected cattle. Many people, including me, found it alarming that these grass-eating creatures were being given food derived from other cows. It smacked of bovine cannibalism and the practice was banned. Nevertheless, the export of British beef was outlawed by the EU for ten years.

As if BSE in livestock was not bad enough, word began to spread that it could pass to humans who had eaten infected beef. At first these reports were not taken too seriously and the agriculture minister John Selwyn Gummer was filmed encouraging his four-year-old daughter to eat a British beef-burger. He took a bite himself and pronounced there was nothing wrong with British beef.

But later a connection was established and at the last count 178 people in the UK have died from the human version of BSE, known as Variant Creutzfeldt-Jakob disease. Sales of beef fell by 40 per cent and only began to recover once all potentially infected meat was out of the food chain. Mr Gummer and his daughter suffered no ill effects and for me it was a powerful introduction to the drama and heartbreak that can suddenly descend on British farming.

Junior health minister Edwina Currie sparked off another food crisis when she told ITN: 'Most of the egg production in this country sadly is now infected with salmonella,' thus creating panic among the public and outrage among egg producers. Sales dropped by 60 per cent overnight, so 400 million surplus eggs were destroyed and four million hens slaughtered. I did a last-minute report for *Countryfile* trying to give some reassurance.

Although there was no disputing the fact that cases of salmonella had trebled, the chances of getting it from eating an egg were about one in two million and an egg was perfectly safe as long as it was cooked properly. Mrs Currie was accused of 'over-egging' the problem, the government paid compensation to the producers and anti-salmonella vaccination for hens was introduced. Prime Minister Thatcher said: 'I had eggs for breakfast' and Mrs Currie resigned fourteen days after making her comments.

Bovine TB is a dreaded and expensive cloud that has hung over British cattle farms since the 1970s and refuses to drift away. It costs taxpayers around £100 million a year in compensation to farmers whose animals have to be destroyed and it is now also costing the lives of thousands of badgers who, in a highly controversial government initiative, are being culled in an effort to control the disease. Cattle often contract BSE in areas where there are large populations of badgers, but where does the blame lie? Do badgers pass it to cattle, is it vice versa, or simply a vicious circle? We've had some emotion-packed reports on the show over the years.

In February 2001, signs of foot-and-mouth disease were found among pigs in an abattoir in Essex, and by the time the last case was confirmed in September six million sheep, cattle and pigs had been slaughtered across the nation. It was the most horrendous eight months in the history of British livestock farming. Exclusion zones were set up around affected farms, huge bonfires of carcasses lit up the skies and farming families openly wept as the men with guns arrived and herds and flocks that had often taken generations to build up disappeared in the flames.

Countryfile had a duty to go live every Sunday to bring our audience, especially the farmers, up-to-date. We had graphic reports from stricken areas and our first was from Warwickshire, where a farmer was in tears as he told me how a team from the ministry had arrived and dispatched his herd as they lay in their stalls. I had to wear protective clothing and we talked across an official barrier erected at the gates.

He bravely agreed to take video footage of the scene. Most of his shots were too gruesome to show on family television but to me, watching them in our editing suite, they illustrated the true and terrible dimensions of this crisis.

Later we included a weekly video report from a Devon

farmer, Paula Walton, as the disease came ever closer to her animals. She captured quite brilliantly the fear, the tension and the feeling of helplessness and isolation shared by so many farmers around the country. Mercifully, her cattle were spared.

Because much of the countryside was closed off, it was hard for farmers, especially isolated older ones who didn't have computers and internet access, to keep fully abreast of the outbreak. So in Cumbria a friend of mine, Ann Risman, set up a project to give them second-hand computers and show them how to use both them and the web. It certainly improved their state of mind and helped them keep in touch with each other through those dreadful days.

The government was accused of acting too slowly as the bodies piled up. It was left to a retired brigadier, Alex Birtwistle, to find a solution to the disposal problem in Cumbria, the worst affected area – and he worked out his plan, he told me, on the back of a cigarette packet. The brigadier decided a mass burial site was needed and he chose a disused airfield at Great Orton near Carlisle. 'I hope it's going to work,' he said, 'because if it doesn't the consequences will be terrible.' It did work. Lorries brought more than half a million carcasses there – it was a horrifying place and the images of those pits and the piles of dead animals will never be forgotten.

When Britain was finally declared free of foot-and-mouth almost a year after it was first reported, farmers whose lives had been devastated set about rebuilding them. And at Great Orton conservationists took up the challenge of turning the burial ground into a place of peace and beauty. It is now the Watchtree Nature Reserve, a 205-acre haven for both wildlife and people, with flower meadows, woods and wetlands.

At this now-beautiful location I had the honour of joining

Alex Birtwistle in planting an oak tree to commemorate the tenth anniversary of that appalling chapter in British agricultural history. It was a sombre yet optimistic occasion in the company of farmers whose herds and flocks had been unceremoniously dumped there and naturalists thrilled to see great crested newts and dragonflies now thriving in what had been a hellhole.

The wild creatures of our countryside feature as prominently (if not more prominently) as domesticated ones on *Countryfile* and often it is because of the threats that face them. There are half as many hedgehogs now as there were in 2000. One in five British mammal species is at risk of extinction, including red squirrels and wildcats.

Farmland bird numbers have dropped by more than half over the last fifty years (that is about forty-four million fewer) and grey partridges, corn buntings, tree sparrows, skylarks, lapwings and nightingales are among those in serious trouble. I count myself lucky to have seen all of them, even the elusive nightingale.

At dawn one morning I was filming with an expert ornithologist and we heard one singing its heart out. He managed to spot it on the branch of a tree, hidden amongst the leaves. Our cameraman focussed in on a little, rather plain-looking brown bird – but what a strong and glorious voice it had.

Blame for the plight of our wildlife is laid at industrial-scale farming, pesticides, habitat loss, disease and climate change. It is, of course, possible to limit the impact and we must ask ourselves: what will future generations think of us if we allow many of the sights and sounds once common in our countryside to vanish?

Fortunately action is being taken and *Countryfile* works

with organizations like The Wildlife Trusts, the RSPB, the Woodlands Trust and many other conservation groups, as well as farming interests, to highlight concerns and show what can be done to make things better.

So not all is gloom and doom and I'm proud to have played a tiny part in one of the greatest conservation successes of recent times. Red kites were once known as dustbin birds because in medieval times they cleaned the streets of food waste. But they were persecuted to extinction more than a hundred years ago except for a few who survived in a corner of Wales.

In 1989 a five-year plan was launched to reintroduce red kites to England at a spot called Christmas Common in the Chiltern Hills. They would be brought over from Spain where the species was plentiful and for *Countryfile* I flew to Madrid to collect one of them in its cage.

British Airways allowed the kite to travel in the seat next to me and fellow passengers were bemused to hear a rather pathetic little squawk – not what you would expect from a mighty raptor. Back at Christmas Common it was kept in quarantine with a handful of others until it was time to release them.

We filmed as their huge wings freed them from the cages and took them up to welcoming skies. They soared effortlessly for a while then landed in nearby treetops. A little cheer went up and everyone involved in the project prayed that these magnificent birds with their distinctive forked tails would thrive once again in the Buckinghamshire landscape.

They did more than that – their descendants can now be found in many parts of the British Isles with getting on for 2,000 breeding pairs. When I see a red kite flying over my garden I wonder if its related to the one who sat next to me on that plane.

*

For most of the 1990s television production at Pebble Mill was dominated by a man called John King, the father of the wildlife photographer and broadcaster Simon King. John reminded me of an old-style actor-manager: flamboyant, risk-taking, and larger-than-life. Among other things, he was in charge of *Going For a Song* and *Tellyaddicts*. He was also *Countryfile*'s boss and he decided we should make the most of our audience's obvious love for the countryside and get them to send in favourite photographs they had taken.

So was born the *Countryfile* photographic competition, and we were inundated with entries, many of an incredibly high standard. Every November we staged a live show and winners in various categories came to the studios in Birmingham to receive their awards. Viewers were invited to take part in a telephone vote to pick the best and the response was amazing – up to 50,000 calls in just fifteen minutes and the production team had only three minutes during the weather forecast to work out the results and pass them to me.

But one year we failed to get on the air. The day before, a massive snowfall had brought most of the country to a halt; roads were blocked, trains were stranded, power lines were down. Finalists and the *Countryfile* studio crew did their best to reach Pebble Mill but most of us failed. I had to abandon my car on the A5 about thirty miles away because the snow was so deep.

I trudged for a couple of miles, past many other abandoned vehicles, until I reached a large, roadhouse-style hotel. In the foyer others who'd been stranded were claiming their sleeping places on the floor as all the rooms had been taken long ago. I was looking for a suitable spot myself when a lovely lady came up to me and said: 'Would you like to join us, John? We're having our firm's annual dinner but there are spare places because some of our team haven't been able to make it here.'

Would I just – she was like an angel appearing with glad tidings. Many of the hotel staff hadn't been able to get to work so the meal was limited but the kitchen staff who were there did a great job and I 'sang for my supper' by giving a little speech afterwards. Then the sales manager asked me: 'Where are you going to sleep, John?' I told him it would be the floor like everyone else, if I could find a space now.

'Look,' he said, 'I have a twin bedroom and my wife didn't set off from home because of the weather. You're more than welcome to her bed – and I won't be getting much sleep tonight anyway because I'll be with my sales team.' I thanked him profusely and, tongue in cheek, said: 'Try not to make too much noise when you come in.'

Next morning I looked out of the window and the snow was still deep – no sign of a thaw. My new-found friend was fast asleep in the next bed so I dressed quietly and made a cup of tea. He awoke, bleary-eyed, at the sound of the kettle and I made him a cup as well.

He sat up in bed as I handed it to him and said: 'You know, John, this is one thing I won't be able to boast about!'

I checked in with the office and was told the show had been postponed until the following Sunday so I made my way back to the car and started digging it out of the drift (on *Countryfile* you always have a shovel in the boot of the car as soon as winter starts). On the radio I heard just how bad the snow had been, but conditions were slowly improving and I managed to travel the relatively short distance home by late that night. Next weekend the weather was lovely and we had a bunch of belated winners.

After a few years we decided we should not let the very best photos disappear from sight when the competition was over and so, in 2001, the *Countryfile* Calendar first appeared. It has become one of the country's best-selling calendars and brightens up many thousands of kitchen walls, while at the same time raising money for BBC Children in Need.

The total raised to date is more than £17 million, and to complete the circle started by his father, Simon King has been one of the judges in recent years.

One regular judge in the early days was the photographer Patrick Lichfield, officially the Earl of Lichfield, cousin of the Queen and a wonderful raconteur. Though he was famous for his images of beauty and style from the swinging sixties onwards he was also passionate about the countryside and its wildlife and I remember sitting with him on a hill where the M40 cuts through the Chilterns as he mused about the impact of modern infrastructures on ancient landscapes. Patrick brought a distinctive, perceptive eye to our judging and off-camera he revealed to me some of his behind-the-lens secrets:

> I was one of the first professional photographers to go digital because it saved me a fortune in film stock and when I was shooting glamour calendars I didn't have to take the models to exotic locations where they would complain non-stop about the discomforts. Instead I could put them in a comfortable studio and impose the images behind them. Sand between the toes was always a problem, though. One top model was supposed to be in a wild desert so I filmed her in a children's sandpit – much less trouble.

Patrick took the formal wedding photographs of the Prince and Princess of Wales at Buckingham Palace and loaded an extra roll of film for the US media. A dispatch rider whisked it to Heathrow where it was put on board a New-York bound Concorde. Within a few hours it was being processed over there, which was pretty quick for 1981. But, he added:

> Contrast that to 1999 and the first time Prince Charles and Camilla Parker Bowles were seen together in public, outside the Ritz Hotel in London. Dozens of photographers were there with digital cameras and news desks around the world saw those pictures within minutes, even seconds. Digital has changed everything.

Patrick told me about a trick he played on a world-famous model he was on location with:

> I asked her if I could take just one shot of her holding a greetings message to send to someone I knew on his birthday. I assured her it would not be published and I had spoken to her agent about fees. She agreed and after I had taken the picture I handed her a one-dollar bill. She was renowned for her quick temper and hit the roof. 'But I know how much you charge,' I called after her, 'and that shot was taken at 1/250th of a second, so I reckon a dollar is about right.' I think she forgave me.

Patrick died from a stroke in 2005 and I felt that the world he had captured so brilliantly with his lens was a duller place without him.

During my first decade on *Countryfile* my most regular on-screen colleague was Rupert Segar, a warm and friendly man who looked more like a ruddy-faced farmer than anyone else on the show. Rupert came to the show from *Newsnight* and did many of the serious reports but he says that among the few images on the internet of his time on *Countryfile* are ones of him having fun faking a crop circle.

At the time there were claims from UFO-believers that aliens and their spaceships were creating strange patterns in the middle of large arable fields and Rupert, having assessed the 'evidence' set about to disprove the claims with the help of the late Doug Bower, a crop circle hoaxer.

A group of believers got wind of their plan, but Rupert and Doug led them a merry chase down country lanes and managed to shake them off. 'We had booked a farmer's field for our demonstration and Doug had worked out an intricate design but unfortunately he confused the radius and diameter' says Rupert who, appropriately, is now a sci-fi novelist and maths teacher.

'It meant the rope we were using to create the circle in the crop was twice as long as it should have been. Doug was fit, fitter than me, but even he was sweating with the effort. We only managed to finish the circle when another bunch of fakers turned up and helped us out.'

The result did look enormously impressive in every sense and confirmed my belief that such installations could never be the work of creatures from outer space. On another memorable occasion Rupert fell off an ox. 'The farmer told me he rode it regularly, but I was halfway on when the soundman moved his microphone and to the ox the furry cylinder at the end of a long stick must have looked like a flying wolf. It reared up and ran off, leaving me on the ground.'

Falling off is a hazard of life for *Countryfile* presenters and so is falling over, especially if you are walking across stony ground or through undergrowth and trying to look at the camera rather than where your feet are taking you. Another colleague in the early days, Michael Collie, managed to fall over no less than eight times while attempting to report from a stretch of bouncy moorland. It looked hilarious on the *Auntie's Bloomers* show.

Rupert says of his time on *Countryfile*: 'Being a small, self-contained production unit it became like family. There were squabbles as well as good friendships and for me the greatest friendship was with the woman I married.' He and Jane Fletcher, one of our producers, fell in love over the camera.

Jane and I worked together on some hard-hitting stories. We were among the first to reveal concerns about the over-use of antibiotics in farming and the danger that resistance to them could be passed though the food chain. We also exposed the persecution of birds of prey, especially hen harriers on the grouse moors. Jane recalls: 'The RSPB were working with big shooting estates in Scotland to try to increase numbers, which had reached critical levels, but the birds kept disappearing or being found poisoned. I still hear the same story today, after all these years, which I find tragic.'

After leaving *Countryfile*, Jane went on to produce *The Sky At Night* for ten years and achieved the distinction of being the only woman, apart from his late mother, who could tell Sir Patrick Moore what to do.

As well as reporting the country life, I was living it with my family in a picture-postcard village called The Lee high on the Chiltern Hills in Buckinghamshire – so high that we were told that if we could see forever the next highest land in our sights to the east would be the Russian Steppes. We had moved there from London in 1981 and our home was in a section of a house once owned by Sir Arthur Lasenby

Liberty, who founded the Liberty's store on Regent Street in London.

It was said to have a ghost, and though none of us ever saw it, friends who stayed overnight occasionally said they sensed something strange in the guest bedroom. Much later Emma and Victoria admitted to me that, if Marilyn and I were out for the evening, they armed themselves with tennis racquets just in case.

We had a busy social life and our local, the Cock and Rabbit, was (perhaps too conveniently) just across the road. Among our neighbours and friends were the actor Geoffrey Palmer and his wife Sally. Marilyn and Sally were on a committee set up to raise funds for the upkeep of the small medieval church behind the village's main church, which was used for social gathering and concerts

Geoffrey (just as funny and lugubrious in the flesh as he is on screen) starred with Dame Judi Dench in that very popular comedy series *As Time Goes By*, and he persuaded her and her late husband Michael Williams to give a 'recital' in aid of the old church. It was an enchanted evening, brimming with memorable poetry and prose, and what really springs to mind is a little couplet beautifully delivered by Michael:

> The task they said could not be done, but he wouldn't listen to it.
> He took on that task that could not be done . . . And he couldn't do it.

A perfect exposition of both arrogance and comeuppance.

I met Dame Judi again more recently when *Countryfile* produced a special edition to mark the 400th anniversary of the birth of William Shakespeare. As well as being one of our finest and best-loved actors, she is also a fan of the show. We walked along a route through the Kent countryside that the Bard's strolling players would have used and ended up in the little town of Fordwich on the River Stour, where the medieval town hall still has a document from the early 1600s proving they had actually performed there – and been paid.

Dame Judi was as lovely and as chatty as I'd hoped she would be and as we stood, literally, in the footsteps of those original Shakespearian actors, I asked her if she could give us just a few memorable lines, maybe the beginning of the balcony scene from *Romeo and Juliet*. Which she did immediately and completely from memory in that little timber-beamed room:

> O Romeo, Romeo! wherefore art thou Romeo?
> Deny thy father, and refuse thy name.
> Or, if thou wilt not, be but sworn my love,
> And I'll no longer be a Capulet.

What she didn't know was that I knew Romeo's response:

> Shall I hear more, or shall I speak at this?

She was taken completely by surprise and burst into delighted laughter. It was a big moment for me, playing out a little scene with an Oscar-winning star and with the greatest respect to Dame Judi, we must have been the oldest Romeo and Juliet ever.

We had other actors visit The Lee, because the pilot episode of *Midsomer Murders* was filmed in and around the village, including in our house. While I was working in my office, John Nettles, as Chief Inspector Barnaby, was in our kitchen investigating some dastardly deed.

And later, when I looked out of the window, there was a team of maybe sixty actors and technicians eating lunch from a catering van. What a contrast to our four-person teams on *Countryfile*, looking for a pub (or maybe just eating sandwiches) for their break. Though Dame Judi must be an expert on location catering, when she filmed with me on the Shakespeare shoot she was quite happy with a chocolate biscuit and a flask of tea from the back of the Discovery, spurning the china cups the team had brought especially for her.

After that the house made several more appearances on *Midsomer Murders* but the best deal came when the BBC arrived to film a 1920s-style detective series. It involved Dame Diana Rigg driving through the village in a period car, and although all the houses were more than old enough their television aerials gave the game away. So we residents were offered a small amount to have them taken down for a few days. When the technician arrived, I asked him if ours would still work in the rather large attic. 'I'll give it a try, mate,' he said, and it did. Not only did we get a fee, we got a roofline free from modern trappings.

Countryfile filmed in The Lee when I made a series of features with the naturalist and writer Richard Mabey based on his *Flora Britannica*, a guide to the social history of our plants and trees. Richard also wrote that classic book about foraging, *Food for Free*, and we wandered the lanes around the village picking wild garlic, mushrooms and other edible hedgerow plants and herbs. Then we took them to the Cock and Rabbit, where the landlord/chef Franco used them to create a fine lunch.

One of the perks of the job is to taste the amazing range of food and drink produced in these islands – purely, of course, to inform our audience of its quality. I've caught and eaten langoustines on the Isle of Skye (the rest were put into seawater containers and shipped to restaurants in the South of France), helped make black pudding from bowls of blood in a kitchen on the Western Isles (just like a horror movie), baked sourdough bread on a campfire (even the burnt bits tasted good), sampled squirrel stew (rather like chicken) and wet my whistle with gin made from ancient rye grains and 'wild' beer brewed from leaves.

It's all in a day's work, and often something of a challenge, none more so than when I had just thirty minutes to eat a five-course, Michelin-starred meal. I was making a film in Paris with my good friend Patrick Flavelle (who climbed the ladder on *Countryfile* from researcher to series producer, a rare achievement) about Label Rouge, the French symbol of food quality. It's awarded to products judged superior to their rivals and Scottish salmon had just become the first fish and the first non-French product to receive the accolade.

We got permission to film in a top Parisian restaurant, La Tour d'Argent, with views from its wide windows of the River Seine and Notre Dame cathedral. Scottish salmon was one of its speciality dishes but we had only a short time to get shots of it being prepared before customers arrived for lunch. The kitchen aromas were fantastic but we couldn't dally – the schedule was tight so, still salivating, we headed to the local fresh food market to film staff from the restaurant buying supplies. Then it would be a dash for our flight home with baguette, perhaps, in hand.

But while we were capturing the market activity the restaurant's owner phoned, inviting Patrick, the crew and me back to La Tour d'Argent as his guests. It was a wonderful, totally unexpected gesture (except in our dreams) but there was only half an hour to spare before leaving for Charles de Gaulle Airport.

It took us maybe a couple of seconds at the most to agree it would be discourteous to turn down such an offer, so we raced back to the restaurant. In what must have appeared an unseemly rush to other diners, who were savouring every morsel, we managed to get through most of the five lavish courses, including the salmon, before our departure deadline. Never has such a magnificent and expensive lunch disappeared so quickly and, alas, there was no time to even glance through the forty-page wine list.

Later I wondered what Ernest Hemingway and Marcel Proust, once regulars at La Tour d'Argent, would have made of the hurried scene. We thanked the owner, who seemed unfazed at the sight of his culinary masterpieces being savaged by a bunch of hungry Englishmen, and caught the plane home – just.

Not long afterwards, Patrick and I found ourselves at the other end of the *Countryfile* 'glamour' scale, in a farmyard filming a report about a scheme to shred used *Yellow Pages* into livestock bedding. We spotted that a cow in the adjacent field seemed in distress, bellowing and pacing, and when we went over to check her we saw the back legs of a calf sticking out. She was giving birth.

We ran to tell the farmer's wife, who tried in vain to contact her husband. When we got back to the yard, the cow had dropped to the ground in obvious pain and we had no option but to do a spot of midwifery. After a great deal of heavy tugging, out popped the calf, seemingly fine. A few minutes later it was on its feet, and we carried on with our report on 'shredding for bedding'.

21. *Prime Time*

Not a lot of people know (to use a famous phrase associated with my first-ever star interviewee, Sir Michael Caine) that moving to Sunday evenings in 2009 was not the first time *Countryfile* had been shown at prime time. Back in the spring of 1994 we had a short run at 7.30 p.m. on Wednesday evenings with a revised edition being transmitted in our Sunday slot.

We put together what I thought was a very watchable series with reports on fox hunting, rural homelessness, badger baiting, the dumping of harmful chemicals – and we even blew up a Skoda. In a recent IRA attack in London the bombers had used the agricultural fertilizer ammonium nitrate in their explosives and caused massive damage.

We showed how easy it was to make such bombs (without, of course, giving away the formula). We got hold of some fertilizer, bought a clapped-out orange Skoda for £50, put it in the middle of a field and stood well back while an explosives expert turned it into a bomb. The result was spectacular – the car was blown to smithereens with wreckage scattered all across the field. Imagine what the impact would have been like in a city street. Our film proved that the sale of these fertilizers needed to be carefully monitored.

Another of my reports examined the fate of many racehorses that didn't live up to expectations. We were told that for hundreds every year there was an inglorious end, being dispatched with a bullet or sold into the pet meat trade. I filmed at a rehabilitation centre in the Lake District where some of the lucky ones were being trained to live a different life with caring owners away from the racetracks.

Our team had high hopes that we would move permanently to the evening schedule – after all, our first outing got more than seven million viewers and reached No. 16 in the weekly audience ratings – but it was not to be.

Though our mailbag was huge and our audience figures continued to be substantial, even when we found ourselves up against *Coronation Street* (it seemed to me to be the perfect alternative to life in Weatherfield), for some reason which was never properly explained to us, *Countryfile* in prime time was not recommissioned.

John King broke the bad news to us shortly before a party we were holding to celebrate the end of the run, and nobody felt like celebrating. One insider in London told me the programme planners wanted 'something a little lighter' in that slot, but the truth is that we could have been one of the BBC's top shows fifteen years before that actually happened.

The man now known as 'Britain's favourite farmer' joined *Countryfile* in 2001 in a rather unusual way – he won a competition. When the programme revealed on air that it was searching for a new presenter and asked viewers to send in videotapes of themselves talking about their love for the countryside, Adam Henson decided to have a go.

From around 3,500 viewers who sent in audition tapes he found himself on a small shortlist of hopefuls invited to spend a weekend with *Countryfile* directors, making films to test their strengths. Adam was the stand-out winner and I went with a camera team to his Cotswold farm to surprise him with the news. We have been friends ever since and he calls me his 'other dad'. His late father, Joe, founded the Rare Breeds Survival Trust and Adam carries on the mission his father started at the Cotswold Farm Park near Stow-on-the Wold in Gloucestershire, where many threatened species of farm animals are in safekeeping.

For years now a visit to Adam's farm has been a weekly treat on *Countryfile* and I think that, almost single-handedly, he has helped to break down the barriers between farmers and consumers. His obvious love for his animals, his passion for the land, his belief in British agriculture and his natural talent as a presenter allow our viewers, especially the urban ones, to discover the ups and downs of farming life. And all because he won a competition.

Adam, or rather one of his oxen, was responsible for one of my most embarrassing television moments. I was leading it calmly across a field when suddenly this huge animal decided it wanted to break free and dragged me for some distance until Adam, skilled animal handler that he is, took control. In fact, I don't have much luck with cattle. I've been upended by a Highland bullock that got its horns under my anorak (no harm done, fortunately), and when I was interviewing a vet while she was examining the back end of a cow it evacuated all over us.

Joining around the same time as Adam was another newcomer to television who had just made a name for himself on a reality show set on the remote Hebridean island of Taransay. Ben Fogle is the one that everyone remembers from the thirty members of the public who spent a year fending for themselves in *Castaway 2000*.

'I loved working on *Countryfile* – it was the happiest eight years in television,' says Ben. 'It was the first show I was offered after *Castaway* and to be honest I had never heard of it. But I grew up watching *Newsround* so I had heard of John Craven.

'In the end I made more than 350 films for *Countryfile* and I can remember them all. The first was about dog sledding in Aviemore. The producer, Teresa Bogan, told me: "Just be yourself." I have used that mantra ever since. My favourites were when I took part in eccentric adventure races across the countryside, from Man Versus Horse to the world worm-charming championships. I also did films about stinging-nettle eating, lawn-mower racing and tin-bath racing.'

In fact Ben made headlines when he competed in the World Tin Bath Championships on the Isle of Man. He had just rowed across the Atlantic with James Cracknell so this must have seemed like a doddle. But producer Jo Brame got a call from the other side of the harbour saying: 'Get over here quick. Ben's in the back of an ambulance with hypothermia!' Fortunately it was only a mild case and it certainly didn't stop him carrying on his adventuring.

Charlotte Smith and Michaela Strachan had joined the show before Ben and Adam, adding their specialist authority in news and wildlife respectively to the programme but covering general stories as well and having fun. Charlotte made a film about farmers stripping off for a charity calendar (inspired by the *Calendar Girls* film) and felt she should get into the swing of things. She ended up wearing just her pants and wellies while holding a board in front of her proclaiming *Countryfile*. Michaela went to report on the World Gurning

Championships in Cumbria, where contestants contort their faces – and won!

Miriam O'Reilly became a Countryfiler when she compiled and presented all the latest news stories in our live programmes during the foot-and-mouth outbreak and we did many shows together, including a 16,000 miles round-trip to the Falkland Islands to mark the 25th anniversary of the invasion.

Considering that Argentina calls the islands the Malvinas – something the Falklanders hotly dispute – we were surprised to discover that in Port Stanley we were staying at the Malvinas Hotel. To me the countryside looked rather like the Yorkshire Dales or the Highlands so it seemed an awfully long way to travel to feel that you hadn't left home.

Miriam talked to islanders about their experiences during the conflict and how they felt now while I took a scheduled flight to some of the outer islands. I couldn't have gone in secret because to my surprise the passenger list was read out on the local radio station the night before. On the islands we filmed two rare species of bird, black-browed albatrosses and a raptor called the cara-cara – and one cara-cara risked being even rarer when it cheekily pecked at my bootlaces and then stole my packed lunch.

I also made a film about the many shipwrecks around the Falklands but the most famous of them all, the SS *Great Britain*, was no longer there. Back in 1970 I had watched in awe as Isambard Kingdom Brunel's iron steamship was towed up the River Avon after being hauled from the depths and transported across the Atlantic. She was returned to the dry dock where she had been built 127 years before and after years of restoration work she is now a fine museum and a fitting memorial to one of our greatest engineers.

Talking of rare birds, as I was a moment ago, reminds me of an ornithological joke I was once told by the late, great Irish comedian Frank Carson, whose catchphrase was 'It's the way I tell 'em'. We were staying at the same hotel – I was on location and he had just finished his stand-up show – and we were having a couple of drinks. Frank told me he was a *Countryfile* fan but he added: 'There's not a lot of fun in conservation is there – not many jokes about it.'

He then proceeded with this one, which I will do my best to repeat:

> This man was up in court for killing a long-faced, flat-eared owl, one of Britain's rarest creatures.
>
> The magistrate said: 'It's a dreadful thing you have done. Do you have any excuse?'
>
> To which the man said: 'There's no money coming into my house, I have young children and, rather than see them starve, I go out with my shotgun in search of something to eat. I spotted this large owl and my children did not go to bed hungry that night.'
>
> So the magistrate said: 'Well, in view of that I won't send you to prison. Instead, you're on probation and don't do anything like this again.'
>
> Later the magistrate saw the man leaving the probation office and said: 'One thing intrigues me. What did it taste like, this long-faced, flat-eared owl?'
>
> The reply came back: 'A bit like a golden eagle.'

After that, you might be thinking that Frank was right, it is the way you tell them, but a few more good conservation jokes wouldn't go amiss.

On another occasion, in another bar, this time on the Isles of Scilly, Adam Henson and I were having a quiet drink after a day's filming and I noticed a man looking intensely at me. After a while he came over and said: 'You don't know me, John, but I once kept very close watch on you. I was in the SAS and we were on a training mission in Slovakia.'

I had been there making a series of films about how eastern European economies were coping after the death of communism, but found it hard to believe that our elite special forces could ever be bothered about my movements. So I asked him to explain: 'Our captain told me he had seen you in the reception area of the Forum Hotel in Bratislava and you were to be our target for the next few days,' he told us. 'It was an exercise and he wanted reports on your every move.'

This seemed so surreal I wondered whether my purported 'shadow' had been drinking a little too much. Could he offer any kind of proof to back up his story? 'Well,' he said, 'if I remember correctly you went to a nuclear power station but they wouldn't let you in.' Now that did happen – we tried to follow up rumours, which turned out to be false, that one of the reactors had a crack in it – and no one else knew about our attempted visit because we did not feature it in the televised report. So our man was telling the truth – I had been secretly followed by the sleuths of the SAS.

Years later, Marilyn and I were on a river cruise along the Danube and called in to Bratislava. The only people following me then (as far as I could see) were fellow passengers on an escorted tour. Nearly all of them were American but word had spread that I worked on a television programme in the UK.

A little later in the cruise a Texan wearing a big stetson approached me and said: 'I hear you do a TV show about cows!' 'A little more than cows,' I told him. 'We also have sheep and pigs and a few people, too.'

On a summer's day in Gloucestershire I was stretched out in a flower-strewn hay meadow and lying alongside me was none other than Jilly Cooper, famous for her racy blockbusters. We were filming a sequence for a *Countryfile* profile on her and the director thought that the two of us should relax in the long grass rather than have a conventional set-up for the interview.

Jilly was up for it and we had a delightful chat about her love for the countryside – she has a charming house in the Cotswolds – and her latest sex-and-saddles bestseller. We had met several times before, most memorably when I was chairing Radio 4's *Start the Week* programme and Jilly was given a rough time about her writing style by another contributor, Kenneth Robinson, who was renowned for upsetting guests. She defended herself well but what the radio audience did not know was that she was very close to tears.

Our *Countryfile* encounter was much more pleasant, certainly not confrontational, as the bees buzzed around us and the world seemed wonderfully calm and peaceful. Afterwards Jilly said she would like to work me into a future novel. I don't think she has (not that I've noticed anyway!) and dread to think what the storyline might be.

Maybe it would involve me covered in foam and surrounded by little plastic ducks in a free-standing bath previously used by Tony Blair. To separate fact from fiction, that really did happen to me when I was shooting links for *Countryfile* at Weston Park, a stately home in Staffordshire where Mr Blair had hosted the G8 Summit in 1998.

Joanna Brame, who is now *Countryfile*'s series producer, thought it would be fun for me to be in the bath, wearing a snorkel and bathing cap, to introduce a film about bog snorkelling. This is a 'sport' where people propel themselves along a soggy trench in a peat bog using flippers and a snorkel.

It's a messy, mucky affair but Charlotte Smith had bravely agreed to have a go and she made it to the finish, triumphantly inviting me to take up the challenge. I declined – a hot, soapy bath in a stately home was much more to my liking, even though it did make me look even more foolish than Charlotte.

Two years after that G8 Summit in Staffordshire Tony Blair set out on a two-day prime-ministerial charm offensive to south-west England in a bid to win over the rural vote at a time when, as he acknowledged, there was a crisis in farming – and I followed him with a *Countryfile* camera team.

Everywhere he went there were angry farmers waving banners, but at one stopping point I saw his charisma in action. A large protest group had gathered outside a farmhouse while the prime minister was inside, shirt-sleeved, meeting local leaders round the kitchen table – a classic Blair tactic. I interviewed some of the protestors, including one who poured scorn on government policy and demanded: 'Why won't he listen to people like us?'

One of Blair's entourage overheard the conversation and a few minutes later the man was unexpectedly invited into the meeting. As he went in he gave me a look which said: 'I'll give him what for!'

Would Tony Blair, with his renowned talent for persuasion, be able to sway this resolute Westcountryman? We waited outside the door, not sure what to expect. As he emerged it was obvious the 'Blair effect' had worked because our man said: 'The Prime Minister is doing his best.'

We had been promised an exclusive interview with Mr Blair at some stage during the two days but it was not until the very end of the tour, as darkness fell and I had almost given up hope, that his communications chief Alastair Campbell said I could have 'a couple of minutes' with him. I needed more than that – there were many serious questions to ask – and the prime minister seemed happy to talk.

Among other things he rejected speculation that a special ministry might be set up to deal with rural problems and said the answer was to co-ordinate across government and work out solutions. In other words, in that well-known phrase of the time, joined-up thinking was needed. As Mr Blair continued, Alastair Campbell stood in my eye-line and kept giving me rather violent wind-up signals, which I ignored – after all, we had waited two days for this interview. Mr Campbell was getting as exasperated as some of the farmers. When we concluded, the prime minister thanked me for an interesting encounter, Mr Campbell indicated quite clearly he was far from happy and off they went into the gathering gloom.

My *Countryfile* interview with another prime minister, David Cameron, twelve years later, was much better planned. We met for an hour at Cogges Manor Farm, a rural heritage centre in his Witney constituency in Oxfordshire. The cameras were set up around the kitchen table and before Mr Cameron arrived a lady who seemed to be in charge of his 'image' wanted to know in which chair he would be sitting. She then checked the camera angles and saw a large Welsh dresser in the background.

'Could we move some of those plates and ornaments,' she said. 'It's all much too fussy.' It just proved that politicians have learned to be careful what is behind them on the screen. An exit sign, for instance, would be the last thing they wanted.

When Mr Cameron came into the kitchen, dressed casually in a jumper – this, after all, was *Countryfile* – the first thing he said to me was, 'I was brought up on you, John!' I don't feel particularly old but it is rather alarming when the man leading the country says you were part of his childhood!

We had a wide-ranging conversation and he had no idea beforehand what the questions might be. I challenged him on his plans to make his administration the greenest government ever (which didn't really happen) and on threats to the landscape from overdevelopment. 'I care deeply about our countryside and our environment,' he told me earnestly. 'I would no more put them at risk than I would my own family.'

Today, I wonder what he would make of the report by the Campaign to Protect Rural England revealing that 15,500 new houses have been approved in areas of outstanding natural beauty in the years since he told me that. We also talked of his plans for a free vote in Parliament on bringing back hunting with hounds (which never happened) and on persuading all other EU countries to enforce farm animal welfare laws as diligently as the UK (still waiting for that).

Frankly, I was impressed by his detailed knowledge of rural issues, even when I pressed him for details, and a few months later when I was invited to a Downing Street lunch for people involved in all aspects of the countryside, he smiled and said he hadn't expected to be grilled by 'a rural Jeremy Paxman'.

We had asked for a few minutes after the interview so the director could take shots of the two of us walking round the farm's garden which we could use to accompany my commentary. I didn't hold out too much hope as he was having a busy day; he'd just come off the phone to Angela Merkel, there were constituency matters to attend to and he was due to give a major speech in Oxford. But he insisted on finding time for our little wander, with no Alastair Campbell figure wagging his finger.

During one of the final weeks of the twentieth century I was driving to a location when my mobile phone rang. I pulled into the verge to hear Marilyn saying: 'Switch off the engine and take a deep breath because I have a very big surprise. Are you ready?' I said I was, not knowing what to expect, and she went on: 'A letter arrived for you this morning from 10 Downing Street saying the prime minister is "minded" to recommend to the Queen that you be made an OBE – an officer of the Order of the British Empire – for services to children's and rural broadcasting.'

Wow – a 'gong'. I could hardly believe it – what an unexpected honour. But (and this really was a challenge for a journalist whose whole professional life has been devoted to passing on information) I had to keep it a secret until the millennium honours were announced on New Year's Eve. It was hard, but my lips did stay sealed.

A few months later Marilyn, Emma and Victoria came with me to Buckingham Palace for the investiture ceremony. Several hundred people were being honoured that day and we gathered together for a briefing from one of the court officials.

He explained that when one's name was called one would walk to where the Queen would be waiting. Her Majesty would pin on the medal, conduct a brief conversation and then shake one's hand. That would be the signal that one's time was over and one would take three steps backwards, turn to the right and exit the royal presence.

It could not have been simpler, but by the time it was my turn I was a bag of nerves. I'm not usually like that before a big event but I certainly was that day. All went well until the Queen shook my hand. Maybe I was feeling overwhelming relief that the encounter had gone smoothly, but instead of taking those three steps backwards I stepped sideways, to the Queen's surprise and my family's embarrassment. Thankfully I quickly recovered and made what I hope was a dignified exit.

Norman Wisdom, one of the great comedians from my childhood, received his knighthood that same morning and deliberately turned his exit into a bit of a performance. Walking away from Her Majesty, he did one of the 'trips' that had made him world-famous and was rewarded with a royal smile.

But it was the television gardener Alan Titchmarsh who got the quote of the day. He said that as he received his medal the Queen told him he gave a lot of ladies a lot of pleasure. Alan later explained that she was in fact referring to the time he had given a talk to Sandringham Women's Institute.

The *Countryfile* team was thrilled when, in 2003, the show doubled its airtime to an hour, still on Sunday mornings. But for me the even bigger thrill of moving to prime time six years later was one tinged with regret. I was proud that after more than twenty years the programme's importance as a conduit of all things rural to a mass audience across the nation had finally been recognized. But the promotion meant farewell to most of the presenters who had not only made the show successful but were also good friends; only Adam and I remained.

The controller of BBC One, Jay Hunt, had no place in the new line-up for Michaela Strachan, Juliet Morris, Charlotte Smith, Miriam O'Reilly or Ben Fogle. Ben remembers:

> We had been constantly told by country folk who couldn't watch the programme because they were working that it should go to a Sunday evening slot and for years we tried to get senior BBC people to listen to us.
>
> I was in New Zealand when I got a call from an executive. 'I have some great news,' he said. '*Countryfile* is going prime time.' I wanted to punch the air I was so excited – we had poured our heart and soul into that show and finally we were being rewarded. Then he continued: 'The bad news is you aren't coming with us.'
>
> My world was shattered. I loved that show. It remains the only time in my life I have been fired from anything. I cried. I wanted to know why. 'You are too inaccessible' came the reply. You can draw your own conclusions as to the meaning but I interpreted it as too white, posh and male. I decided to remain quiet with my disappointment. The show of course flew and I was thrilled for Adam and JC that they got to ride the wave of popularity with our beloved show. I haven't done too badly since, but *Countryfile* will always be in my DNA.

Jay Hunt brought in Matt Baker, former star of *Blue Peter*, and Julia Bradbury, already well-known for her television 'walks', to head the presenting team and I became the programme's investigator, examining a different rural issue every week. Adam, who had no specific role before, found that with the new format he wouldn't have to travel very far – his weekly reports would be from his farm in the Cotswolds.

As for those who didn't make the cut, Michaela was quickly back on the rural scene with *Springwatch* and all the other seasonal 'watches', Ben is a global TV adventurer, Charlotte continued her journalism on Radio 4's *Farming Today* and is also now back on *Countryfile*, Juliet is busy with corporate work and Miriam successfully sued the BBC on the grounds of age discrimination in a case which made headlines and, I believe, changed attitudes in the industry. There is now a larger quota of older women, and men, on our screens, and rightly so, because it could be argued that we represent the largest slice of the audience – most viewers are over the age of fifty.

At the time I found it hard to understand why the changes were necessary but I supposed Jay Hunt was, in that all-embracing media word, wanting to 'refresh' the show. She was certainly taking a gamble placing it at the start of what was often, in audience terms, BBC One's biggest evening of the week.

We would be on air before *The Antiques Road Show* and after that would be a major drama or documentary. I had long hoped for a return to prime time but I thought maybe, if we were lucky, it would be on BBC Two, perhaps on a Thursday. Instead, we got a top weekend slot, so how would we fare? It was nail-biting.

The morning after our first prime-time transmission I was in France filming an investigation into the decline in honey bees caused, so it was claimed, by farmers using neonicotinoid pesticides (a report which, incidentally, won an award) when my mobile rang.

It was Jay from her London office. 'We're all jumping up and down with excitement here,' she said. 'We have just seen the overnight viewing figures and more than six million watched last night. A triumph!'

Matt and Julia were an instant success as *Countryfile* became even more approachable to a bigger, more diverse audience. 'Immersion' was the new watchword and we all became much more involved in the subjects we were covering – we all 'had a go'. Some critics believed that *Countryfile* was dumbing down but I strongly disagreed. While it is vital we discover and respect countryside skills and knowledge and highlight them on the show, viewers get a much better impression of how demanding and complex they can be if one of us tries our hand – and, in my case at least, often makes a mess.

Matt, who grew up on a farm in County Durham, has been a continuing major asset and still finds time to add his enthusiastic touches to almost every edition, even though his incredibly busy broadcasting life also includes *The One Show*, wildlife specials and commentating on international gymnastics. Julia brought her strong, compelling personality to the show and when she left, her wellies were more than adequately filled by Ellie Harrison, a passionate naturalist who is also wonderful with people.

More recently we've welcomed two other brilliant women who don't mind roughing it in the worst of weathers, farmer's daughter Helen Skelton (a graduate of *Newsround*) and

Anita Rani, like me from God's Own County and bringing with her the warmth and curiosity of a city girl wanting to discover everything about country life. More recently the presenting line-up has swelled to include Margherita Taylor, Steve Brown, Sean Fletcher and Joe Crowley – we could form a football team!

I carried out my weekly investigations for four years, looking into everything from mega farms to the closure of rural services (hospitals, schools, bus services, shops, pubs, etc) to the mass exodus to the towns of rural young people who can no longer afford to live in the villages where they were born. Journalistically it was some of the most satisfying work in my whole career but eventually I decided I needed a rather quieter, slower life.

Travelling back and forth across the country for forty-six weeks of the year was taking its toll so I cut down my commitment to around twenty shows. Tom Heap took over the investigations and I concentrated on feature items, and that is how it has been ever since. And I've gone full circle because I also introduce the spin-off show, *Countryfile Diaries*, on daytime television.

I'm often asked how many people it takes to make *Countryfile* and how far ahead of transmission is it filmed (many of us in the business still use the word 'film' even though we should really say 'record'). Well, because it is on the air every Sunday of the year, *Countryfile* is an endless conveyor belt that needs at least forty people to operate it.

We have an editor, a series producer, three producers, fourteen directors, ten researchers, two runners and seven people on the production management desk who do the paperwork, the booking of hotels and transport and everything that keeps that conveyor belt going.

Plus, just down the road from our offices at BBC Bristol (we moved there from Birmingham in 2012) is a freelance company where the editing, dubbing and fine-tuning are done for each programme.

It might seem like a large team but every job is vital and, compared to many other programmes, *Countryfile* runs on a really tight budget. I've been told that if you watch every episode the total cost out of your licence fee (£154.50 at the time of writing) would be less than one penny a year. That's value for money!

Each week the show comes from a different area of the country, so a lot of forward planning is needed to pick the right places that have not only beautiful scenery but also interesting stories to tell. A director and a researcher work on the 'futures' desk and are assigned to look for locations up to four months ahead. The programme itself works on a six-week cycle.

During weeks one and two a researcher, working closely with the producer of that show, finds five stories that can be turned into films for the presenters who will be fronting that particular show. Two directors join the researcher in week three to finesse the stories and travel to the locations, often at the opposite ends of a county, to check out the views and meet the people who've been selected for interview. Those contributors will be briefed on what to expect when the crews arrive while back at base the producers check through the stories and add their thoughts on how they should be shot.

On the Thursday and Friday of week four all the stories and links are filmed by two separate teams no matter what the weather. Unlike some shows with bigger budgets, we can't afford to wait until the rain stops or the wind dies down.

We press on despite what nature throws at us and I think that makes the show more realistic. Everyone knows that, for example, you can't always guarantee beautiful sunshine in the Lake District – there's a valid meteorological reason why the place has got lakes.

One presenter works for both days with one of the teams and, because Matt Baker is doing *The One Show* on Thursdays, he leaps into a car or onto a train or plane as soon as it's over so he can join the other crew on Friday. Myself or another presenter will have already filmed with that team on the Thursday, so no time is wasted.

At the end of shooting the two presenters get together, which sometimes means a long drive across country, to say goodbye to camera and give a brief preview of next Sunday's show. It's often the only time we meet except for *Countryfile Live*, our annual four-day country fair when we get to meet large numbers of the audience in person, and 'fun' things like dressing as country and western singers for *Children in Need* night.

During that same week, another crew will be filming Tom Heap's current affairs spot and Adam Henson will be looking at a farming story. Then everyone heads home through the Friday night traffic – except for Adam, who is usually already on his farm.

During week five the directors check their 'rushes' and then head to the editing suites to put their films together. In week six the entire programme is compiled, the presenters record their commentaries to accompany the pictures and the show is ready for transmission on the Sunday. A space is left in the recording so that the five-day weather forecast can be broadcast live during that gap.

So, each *Countryfile* is six weeks in the making, with just sixteen days between filming and transmission. And when you settle down to watch our latest efforts the next two programmes are already 'in the can' and the *Countryfile* conveyor belt is trundling on.

Much as I have loved treading Britain's green acres all these years, I did suggest to my agent, Jo Carlton, that it would be a rather pleasant change to be indoors for a while in a nice warm studio with a shiny floor and lots of flashing lights. So she looked around and found me a brand-new quiz show-with-a-difference called *Beat the Brain*.

Instead of questions on general knowledge it tested the way your mind works, with lots of quirky memory tests, and our teams ranged from gay rugby players to lady poker players. I thoroughly enjoyed this new experience and it kept me out of the rain for a while. *Beat the Brain* was shown on BBC Two and was co-produced by Syco, Simon Cowell's production company, so as a bit of a bonus I got an invitation to Simon's lavish summer party in the orangery at Kensington Palace.

He certainly picks the right places. 'You're looking really well, John,' he told me. 'How much work have you had done?' 'No Botox for me,' I said, 'just fresh country air.'

For many years I've had a rather important off-screen role on *Countryfile* – Santa Claus at the team's Christmas party. I make a discreet exit after the main course and a few minutes later who should appear but a man in red with lots of presents. A few days beforehand everyone had pulled a name out of a bag, a bit like a sweepstake, and had to buy that person an inexpensive but preferably amusing gift.

After Santa has done his duty and waved goodbye I return

to the party and everyone shouts: 'Where've you been, John? You've missed him!' After that comes a spoof awards ceremony when unsuspecting victims are singled out to receive their 'Cravens' – our take on the Oscars, but not as complimentary!

When I celebrated my twenty-five years on *Countryfile* in 2014, the team gave me a cake – a quite amazing cake. In a field of green icing were some icing sheep, and leaning on an icing gate was an icing me, together with an inscription that read: 'Outstanding in his field'! It was one of the most touching gifts I've ever received – and the sheep, gate and I survived for a long time. But not the cake.

To mark the show's 25th we invited a guest editor to take over – none other than the Prince of Wales. He chose issues close to his heart such as the problems facing upland farmers, better quality school meals and greater access to the countryside for those with disabilities. We met in one of the fields on Home Farm on his Highgrove estate and I asked him if he watched the programme. He replied with that famous *House of Cards* sentence: 'You may think so, but I couldn't possibly comment!' But I know he does because later the Duchess of Cornwall told me they enjoy the show.

We celebrated the 30th anniversary in 2018 with a special edition looking at some of our past highlights while contemplating the future for our countryside. I'm now the sole survivor from the early days and over the years I've worked with more than a dozen series producers, maybe a couple of hundred directors, researchers and back-up staff and the best cameramen and sound recordists in the country.

All of them take huge pride and satisfaction in the job of

bringing our countryside into the homes of our viewers and I can honestly say that, just like the *Newsround* team, they have been a great bunch of friends as well as colleagues.

I don't have any plans to give up television – in fact I once said in an interview that, rather like a trusty old workhorse, I would have to be retired rather than choose to retire. Far too many people I know go rather rusty when they give up the day job. During my career I have been extremely fortunate to work on three groundbreaking programmes which were world firsts – *Newsround*, *Swap Shop* and *Countryfile* – and which, I like to think, have been true to the BBC's Reithian principles: to educate, inform and entertain. Some people in television are remembered for programmes which they personally would prefer to forget. Not me.

I'm proud of my shows and of the people who have worked alongside me to make them successful, and I'm grateful to the millions of viewers who have had faith in me over many years. So I'll sign off with what I've said on what seems like a million occasions on screen – bye for now.

22. And Finally . . .

I thought you might like to be reminded of some of those little stories that we used to say goodnight with on *Newsround*, so that children didn't go to bed and have nightmares.

- In New York a rabbit called Harvey bit at least sixteen people, so a charity organization hired him to guard their offices and a notice was posted to warn would-be intruders about Thugs Bunny.
- From Wellington, capital of New Zealand, came news that seventy police dogs were being issued with rubber boots to protect their paws from broken glass when they were called to violent incidents. Maybe they were Wellington boots.
- An MP said that cows who, by ancient law, were allowed to wander along roads in Epping Forest, should be daubed with fluorescent paint to stop them being a danger to traffic at night. What heifer next?
- Swiss trains always leave on time – except when a pet mynah bird in a cage at one railway station became so expert at imitating whistles that it set off lots of trains. The bird was sold – obviously it had ideas above its station.
- How about buying your own farm the size of London for only £12.50? The snag is, it would be on

the planet Mercury where the temperature is 426 degrees Celsius. Talk about hot property.

- In Athens one night a soldier was guarding four Mirage jet fighters with long tapering nosecones and, getting bored, he decided to swing on one of them. To his horror the nose started to bend and he couldn't straighten it. So he swung on the others as well, hoping it wouldn't be noticed. Teams of technicians fixed the noses and the soldier had to face the consequences.
- A small iceberg weighing a tonne was airlifted from Alaska to a conference 3,000 miles away as a symbol of how the world could make use of them. It cooled the delegates' drinks – the first time ice has cost more than drinks.
- In Australia a man predicted that the town of Adelaide would be destroyed by an earthquake and tidal wave so he moved to the town of Warwick, which he said was the safest place in the country. A month later, Warwick had its worst floods in fifty years.
- In Mannheim in West Germany, eighty-five people with sensitive noses went round town several times a day and were asked to sniff out the worst smells caused by pollution. The hope was that the city would be a sweeter place, but did it work? Who nose.
- President Jimmy Carter's daughter Amy needed some help with her history homework but officials thought it was the President himself who needed the information and a team of highly paid officials spent a weekend digging into the past – it must

have been the most expensive piece of homework in history.

- A lady in Russia celebrated what she said was her 130th birthday and all the seventy guests were over the age of a hundred. The local candlemaker must be a rich man.
- Japan's bullet trains were brought to a halt by a bunch of crows that caused short circuits by building their nests in electricity pylons and cutting off the supply. Railway bosses ordered the pylons to be treated with bird repellents and once again the world's fastest trains had something to crow about.
- In Australia, the tomato crop failed and there was a shortage of everyone's favourite sauce. Next season farmers grew many more tomato plants, hoping to ketch-up.

Acknowledgements

I would like to thank everyone who has helped jog my memory while I have been compiling this memoir, in particular Edward Barnes, Mike Beynon, Jo Brame, Lewis Bronze, John Exelby, Helen Fielding, Patrick Flavelle, Ben Fogle, Brian Hawkins, Nick Heathcote, Michael Murphy, Jill Roach and Nick Robinson. My grateful thanks also to my sister Jean for her memories of our early years; to my wife Marilyn for all her recollections; my daughter Victoria and my agent Jo Carlton for being my 'guardians' during the writing process and to Daniel Bunyard and Beatrix McIntyre at Michael Joseph for being so patient and supportive.

He just wanted a decent book to read ...

Not too much to ask, is it? It was in 1935 when Allen Lane, Managing Director of Bodley Head Publishers, stood on a platform at Exeter railway station looking for something good to read on his journey back to London. His choice was limited to popular magazines and poor-quality paperbacks – the same choice faced every day by the vast majority of readers, few of whom could afford hardbacks. Lane's disappointment and subsequent anger at the range of books generally available led him to found a company – and change the world.

'We believed in the existence in this country of a vast reading public for intelligent books at a low price, and staked everything on it'
Sir Allen Lane, 1902–1970, founder of Penguin Books

The quality paperback had arrived – and not just in bookshops. Lane was adamant that his Penguins should appear in chain stores and tobacconists, and should cost no more than a packet of cigarettes.

Reading habits (and cigarette prices) have changed since 1935, but Penguin still believes in publishing the best books for everybody to enjoy. We still believe that good design costs no more than bad design, and we still believe that quality books published passionately and responsibly make the world a better place.

So wherever you see the little bird – whether it's on a piece of prize-winning literary fiction or a celebrity autobiography, political tour de force or historical masterpiece, a serial-killer thriller, reference book, world classic or a piece of pure escapism – you can bet that it represents the very best that the genre has to offer.

Whatever you like to read – trust Penguin.